LIFE OF MINE CONFERENCE 2023

Exploring the holistic life-cycle of a mine

2–4 AUGUST 2023
BRISBANE, AUSTRALIA

The Australasian Institute of Mining and Metallurgy
Publication Series No 4/2023

AusIMM

Published by:
The Australasian Institute of Mining and Metallurgy
Ground Floor, 204 Lygon Street, Carlton Victoria 3053, Australia

ISBN 978-1-922395-23-8

ORGANISING COMMITTEE AND REVIEWERS

Prof Claire Côte
Conference Organising Committee Chair

Alex Thin
FAusIMM(CP)

Brad Radloff
MAusIMM

A/Prof Glen Corder
FAusIMM(CP)

Ingrid Meek
MAusIMM(CP)

Dr Jason Dunlop

Jeremy Durbin
MAusIMM

Rebecca Wright

Sarah McConnell
MAusIMM

Sibasis Acharya
FAusIMM

Todd Bell
MAusIMM

Vanessa MacDonald

AUSIMM

Julie Allen
Head of Events

Joelle Glenister
Manager, Events

Samara Brown
Conference Program Manager

FOREWORD

Welcome to AusIMM's sixth Life of Mine Conference. After all the disruptions of the last few years, we are delighted to host this conference face-to-face in Brisbane, which will foster stimulating discussions.

Since 2012, Life of Mine has sought to promote holistic thinking across all aspects of a mine's life cycle. This edition further strengthens this vision, with the most extensive program to date including around 45 technical presentations. This reflects the recent growing activity and interest in this space, driven by regulatory landscape changes and associated responses from industry and researchers.

Our fantastic keynote speakers will share their views on collaboration, integration, partnerships and technical innovation. Alongside our keynote addresses we have two panel discussions on Indigenous participation in mining, led by Kia Dowell, and Decarbonisation, led by Professor Neil McIntyre.

Once again, Life of Mine has attracted professionals from all disciplines across industry, consulting, government and non-government organisations, as well as academic and research organisations. A continuing strength of this conference is the participation of AusIMM's traditional core constituencies in geology, planning, mining and mineral processing, alongside those working in environmental, social and broader sustainability-focused roles.

What has changed significantly over the last few years is the increasing influence this latter group has within AusIMM and our sector. With this in mind, we have purposely designed a program to facilitate cross- and inter-disciplinary learning and engagement.

We wish to thank our presenting authors, whose contributions are critical to the success of this event, as well as our abstract reviewers for your time and rigour. The support of our sponsors and exhibitors is also crucial, and we acknowledge your genuine interest and support.

The organising committee has been instrumental in making the conference a reality, and I am extremely grateful for their time, advice and commitment. As my first time as Conference Chair, it has been a pleasure to work closely with the AusIMM team to ensure another successful conference.

On behalf of AusIMM and the Sustainable Minerals Institute at The University of Queensland, along with AusIMM's Southern Queensland Branch and Community and Environment Society, it is my pleasure to welcome you to the Life of Mine Conference and we hope you enjoy the content and networking on offer.

Yours faithfully,

Claire Côte

Sustainable Minerals Institute, The University of Queensland

Life of Mine 2023 Conference Organising Committee Chair

SPONSORS

Major Conference Sponsor

BHP

Platinum Sponsor

GHD

Gold Sponsor

srk consulting

Silver Sponsors

AngloAmerican

Deswik

Welcome Reception Sponsor

WSP

Name Badge and Lanyard Sponsor

AMC
consultants
mine smarter

Technical Session Sponsor

VERACIO

Conference App Sponsor

GLENCORE

Supporting Partners

SIBELCO

stanwell

CONTENTS

Decarbonising mine operations, from concept to closure

Developing regional approaches – shifting from operating independently to collaboratively

Effective regulatory frameworks for mine life and beyond

Embedding the circular economy into life-of-mine planning

Innovative rehabilitation and closure solutions

Integrated sustainability planning

Decarbonising mine operations, from concept to closure

Dynamic influences of optimisation on emissions

P J Bangerter[1] and J Pan[2]

1. FAusIMM, Principal Consultant, Orchardman Pty Ltd, Forest Lake Qld 4078.
   Email: philip@bangerter.net.au
2. Consultant, Whittle Consulting Pty Ltd, Surrey Hills Vic 3127.
   Email: jason@whittleconsulting.com.au

INTRODUCTION – THE STATE OF PLAY

Optimising for Net Present Value (NPV) with a strategic optimiser such as COMET, Prober-E, Blaser or Minemax has hitherto lacked a dynamic method for analysing the so-called third-order outcomes (such as Carbon emissions) of optimisation within life-of-mine planning.

Optimisation, *per se*, has natural effects on carbon emissions in a life-of-mine plan. Contemporary strategic planning employs a mathematical optimiser to make sense of non-linear data (ie a block model) and produces a best-NPV optimised mining schedule, within a set of constraints. The very act of optimisation influences emissions, and in a dynamic way.

FIRST, SECOND AND THIRD ORDER

Strategic Planning can be applied at the first, second and third-order (Bangerter, 2022).

First-order effects concern assembling capital and operating costs and calculating a net-present-cost for these. Financial impacts can be compared across options or scenarios, as required.

Second-order effects concern the orebody as an integrated whole and its optimisation.

Pit or Stope designs can be optimised using appropriately directed Lerchs-Grossman algorithms (Lerchs and Grossmann, 1965). Lerchs-Grossman is effective at determining the economic 3D shapes considering block grades, open pit slopes or underground design characteristics, costs, recoveries, and metal prices. Such an optimised schedule utilising these designs should raise early cut-off where physically possible, to increase early metal production, even if positive margin material is discarded and the mine life shortened (Lane, 1988). Introducing appropriate activity-based costing allows the optimisation to account for real world cost differences between bringing different ore types to product such as haul distances and ore hardness, directing designs and schedules to account for maximising margin rather than just grade and revenue (Whittle, 2023).

A re-optimisation will thus change the mining schedule, have a dynamic cut-off policy and even modify the mining design shapes. Such changes become strongly reflected in the emissions.

In the general case, optimising for NPV with such a strategic optimiser and methodology, will naturally favour the inclusion of shallower, higher-grade, softer material and in consequence reduce emissions per tonne of output (ie improve carbon intensity compared to a poorly optimised schedule).

Additionally, there are further consequences for any cost and price variations tested in the planning and subsequently adopted as the plan. These include:

- For mining assets with processing-dominated energy (such as base metals open pit) lowering costs or a predicting a more elevated commodity price naturally <u>increases</u> carbon intensity. It favours expanded exploitation of existing pit-shells which brings marginal material (lower-grade, harder and more distant) to plant rather than waste.

- Whereas, in mining-dominated energy assets (such as iron ore) cut-offs are lowered, life-of-asset increased, and pit-shapes expanded. Marginal material is thus brought into the schedule, <u>increasing</u> the carbon intensity as well.

- And by contrast, when planning introduces process efficiencies (technology improvements, efficiency drives), higher throughputs are possible for the same energy – this will <u>reduce</u> emissions intensity.

Finally, third-order effects are concerned with environmental and community value or impact and an evolving full assessment of sustainability impacts of each choice between options. These are best illustrated in case studies.

CASE STUDIES

Decision-making in this space can look at emissions reductions/increases as a total and as an intensity; all judged against NPV or other financial indicators. Recent case studies can be used to illustrate the decision-support of such an analysis, including technology comparisons and decarbonisation strategies.

Recently, Nordic Iron Ore (NIO) revisited their pre-feasibility level 2019 strategic planning which now includes both a new carbon model and a decarbonisation strategy that includes fleet electrification for their project Blötberget (Hamerslag, 2022). Figure 1 compares the total estimated emissions for the originally planned diesel-powered fleet and the substantial reductions afforded by electrification in a low-carbon grid (Sweden).

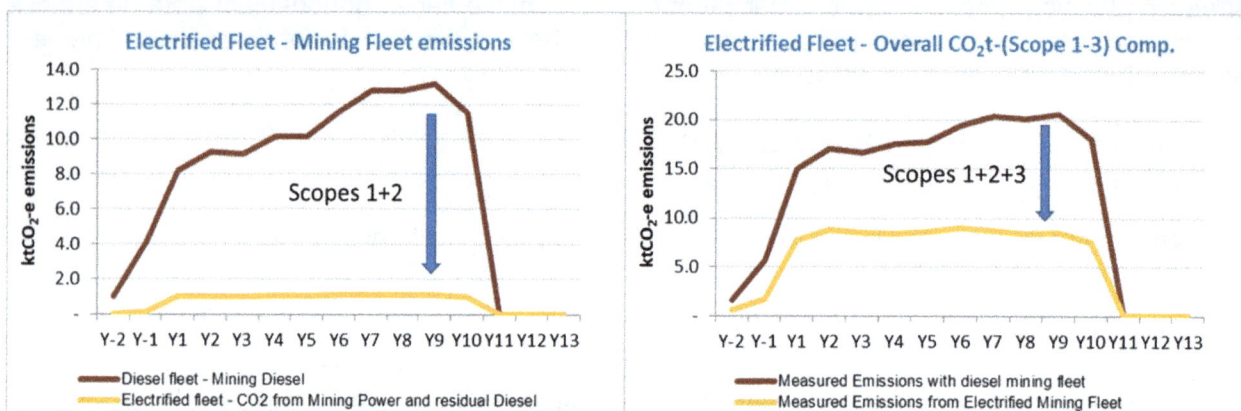

FIG 1 – Nordic Iron Ore Electrification Implications.

Scope 3 emissions include estimates of the material upstream components (such as grinding media, explosives, cement/concrete and transport in the supply chain) as well as downstream transport emissions to the customer's gate. The importance of Scope 3 emissions in several analyses has become apparent and can be further illustrated by the example in Figure 2.

*Carbon emmissions **Scope 1 & 2 only** vs. NPV* *Carbon emmissions **Scope 1, 2 & 3** vs. NPV*

FIG 2 – DPM emissions guiding strategy decisions.

Here, Dundee Precious Metals (DPM) have used their integrated strategic planning study at Chelopech, judging improvements to long-range forecasts (LRF) and hence integrating climate decision-making into capital allocation decisions and transport of concentrates (Nolte, 2022). Assessing the carbon footprint versus the net present value of the operation became a key feature of the study. It found that:

- Scope 1 and 2 emissions are insensitive to LOM scenarios meaning maximisation of NPV does not increase absolute carbon emissions (left graph).

- But that incorporating Scope 3 emissions into the modelling presented some decision challenges in terms of 'upstream' and 'downstream' transport and processing assumptions because Scope 3 emissions are sensitive to the LOM scenario selected – mostly because of the downstream transportation (to alternative destinations other than DPM's Tsumeb operations (DPMT)) and processing of concentrate (right graph).

This reveals a strategic question to be answered: does the choice of customer expose concentrate producers to decarbonisation pressures? In a regional context, this can be yes. A more distant (Asia versus Africa in this case) and more carbon-intensive transport mode will show up in a Scope 3 inventory.

Other examples from recent studies have been:

- fleet electrification with trolley assist

- in-pit crush and convey versus truck haulage

- renewables penetration versus traditional diesel at remote sites

- dry-stack tailings versus conventional tailings storage.

WIDER IMPLICATIONS

Of course, third-order effects are not limited to carbon emissions and any sustainability-related issue can be included. The authors are seeing the positive implications for decision-making of widening this type of dynamic analysis to tailings and water, for example. The trade-off analysis between NPV and tailings volumes can now be routine in strategic planning and is expected to be extended to analyse grind size against not only water recovery, but also water-use intensity and ultimate tailings density. Moreover, it would not be surprising if coarse particle flotation and the emerging Hydraulic Dewatered Stacking methodology (Newman, 2023) would prove to be an interesting third-order analysis in the very near future.

ACKNOWLEDGEMENTS

The authors would like to acknowledge Whittle Consulting for the support afforded to us during the writing of this extended abstract and accompanying presentation material. Moreover, we greatly appreciate the kind permission to use examples from studies conducted for Whittle clients; namely by Mirco Nolte of Dundee Precious Metals and Ronne Hamerslag of Nordic Iron Ore.

REFERENCES

Bangerter, P J, 2022. Implications of ESG on Mine Planning and Equipment Selection, webinar presentation, Djakarta Mining Club, Jakarta.

Hamerslag, R, 2022. Nordic Iron Ore AB, Liam Forum, Brock University, Toronto.

Lane, K F, 1988. *The Economic Definition of Ore* (Mining Journal Books Limited: London).

Lerchs, H and Grossmann, L, 1965. Optimum Design of Open-Pit Mines, *Transactions CIM, LXVII*, pp 17–24.

Newman, P, 2023. Progress on Hydraulic Dewatered Stacking (HDS), retrieved from Anglo American: https://www.angloamerican.com/our-stories/innovation-and-technology/progress-on-hydraulic-dewatered-stacking-hds-el-soldado-chile

Nolte, M, 2022. Holistic Approaches, Integrated Strategic Planning & Sustainable Resource Development, Liam Forum, Brock University: Toronto.

Whittle, G, 2023. Activity Based Costing (ABC) and Theory of Constraints (TOC), retrieved from Whittle Consulting: https://www.whittleconsulting.com.au/integrated-strategic-planning/#activity

The importance and abundance of lichens and mosses on the restored landscape in the nickel-copper city of Sudbury, Ontario, Canada

P J Beckett[1], T Miller[2] and S Wainio[3]

1. Professor Emeritus, Laurentian University, Sudbury Ontario P3A 5B5, Canada.
   Email: pbeckett@laurentian.ca:
2. Student, Laurentian University, Sudbury Ontario P3A 5B5, Canada. Email: tmiller7@live.ca
3. Project Team Leader, Trow Associates Inc., Sudbury Ontario P3E 5M4, Canada.
   Email: shelley.wainio2@ontario.ca

INTRODUCTION

Lichens are a symbiosis of fungi and algae. Air borne elements and gases are absorbed very efficiently over the entire surface of the lichen, because they have no protective structures such as a waxy cuticle or stomata. A survey in 1977 (one year before the Sudbury Regreening program began), indicated that several *Cladonia* spp had invaded the soil in Sudbury's open birch woodlands. Pollution sensitive species such as *Cladonia rangiferina* and *C. mitis* occurred 20–25 km from the smelters. It was anticipated that the improvements in air quality and reclamation efforts would allow a natural invasion of pollution sensitive species into what once was a lichen desert that surrounded the smelters. The purpose of this project was to study the invasion of terricolous (soil-inhabiting) communities of lichens and mosses on reclaimed land in the Sudbury region 45 years after commencement of the Sudbury Regreening Program.

THE SUDBURY LANDSCAPE REGREENING EFFORTS

Cu-Ni mining developed after completion of the transcontinental railway in the mid-1880s. Smelter atmospheric releases of SO_2 were 2.5×10^6 t yr^{-1} in the late 1960s. In 1974 there were atmospheric releases of approximately 5.0×10^4 t yr^{-1} of fine Fe-, Ni-, Cu-, As-, and Pb-containing aerosols. A century of interacting factors, such as logging, wildfires, winds, and water erosion, SO_2 fumigations, acidification, and metal deposition, resulted in 17 000 hectares of highly damaged watersheds and another 64 000 hectares of stunted woodlands. Contaminant reductions commencing in 1972 (98 per cent reduction in SO_2 emissions and large smelter particulates by 2022) created the possibility of re-establishing vegetation.

In 1973 local government formed the Vegetation Enhancement Technical Advisory Committee (now VETAC – Regreening Advisory Panel). VETAC guided the internationally recognised Sudbury Method for technogenic barren landscape restoration. The original Sudbury Method, used from 1978 to 2010, spread 10 t/ha dolomitic limestone of various particle sizes: 400 kg/ha of 5:20:20 fertiliser; a 40 kg ha nursery cover of five to seven agronomic grasses and two legumes; followed by planting of native trees. Over the 45 years of tree and shrub planting Jack Pine (*Pinus banksiana*) and Red Pine (*Pinus resinosa*) account for about two-thirds of the planted trees.

From 2010 regreening activities moved a more complete biodiverse restoration strategy. The current Sudbury Protocol dictates liming, fertilising where needed, and use of native grasses. The planting of a wide array of native trees establishes a forest structure with an emphasis on introducing forest shrubs and understory components (about 80 species). By 2022, 3500 hectares have received limestone, fertiliser, and seed and 10.6 million trees and shrubs have been planted for approximately Can$36.4 million.

METHODS

A chronosequence of reclaimed sites was established that span the years of operation of the Regreening Program and have similar landscape characteristics of gentle hillside slopes, stony substrates, a similar aspect and developing forest of pines. For comparative purposes a comparison site was selected in the pine forest 50 km south of Sudbury. At each site along three transects of 5 m a series of contiguous 20 × 20 cm quadrats, were placed and identification, percent cover of each lichen and moss species determined. Soil samples were also collected at each site and analysed for pH. Samples of *Cladonia rangiferina* were collected for elemental analysis.

RESULTS

Recently reclaimed sites were dominated by grasses and legumes with few shrubs or trees except for small planted conifers. Older reclaimed sites from 1979 to 1992 have been invaded by herbaceous vascular plants. bryophytes and tree species such as Trembling Poplar (*Populus tremuloides*), White Birch (*Betula papyrifera*). Red Pine, (*Pinus resinosa*), Jack Pine (*Pinus banksiana*) and White Pine (*Pinus banksiana*) dominated the sites and gave the appearance of an open woodland. The substrate pH ranged from 4.1 in the unlimed site to 4.6–5.1 in the limed sites.

The total number of lichen and moss species ranged from eight in the recently reclaimed sites to 17 at a site reclaimed over 40 years ago (Figure 1). The oldest reclaimed site had a similar number of species to the forest comparisons site outside Sudbury (Figure 1). There was a gradual increase in number of species with time. Most of the species were members of the pixie cup *Cladonia* lichens.

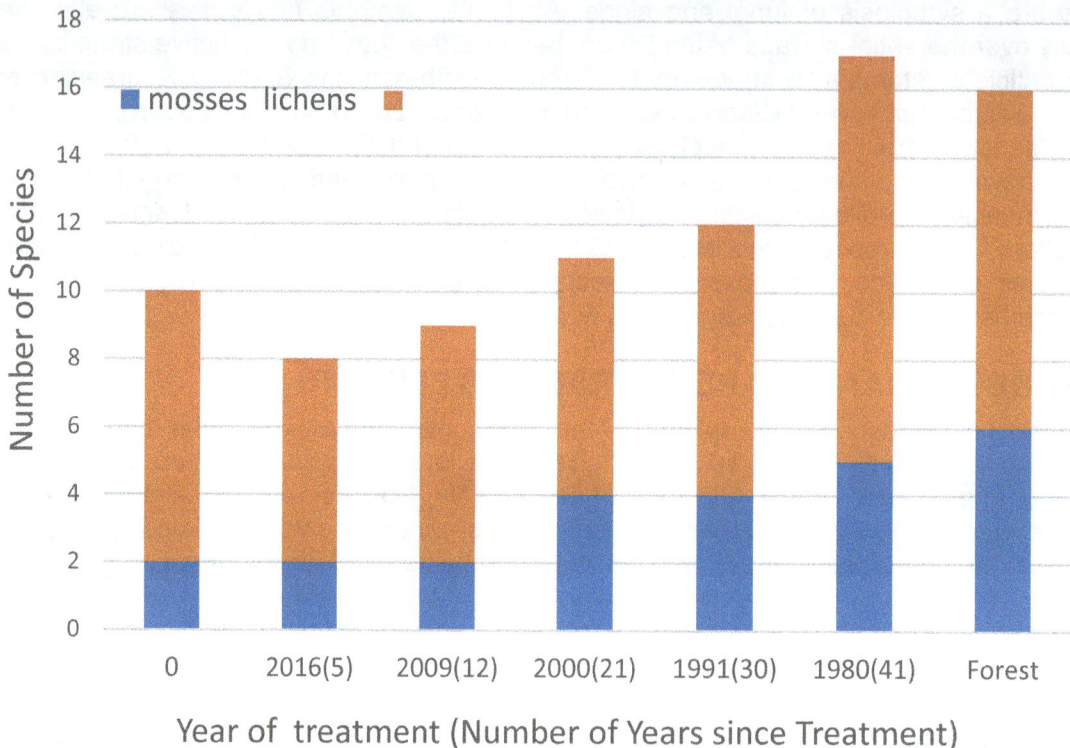

FIG 1 – The abundance of lichens and mosses at a site with no regreening treatment (0) and a series of regreened sites with age of year of reclamation (and number of years reclaimed dating back from 2021).

Cladonia cristella, C. pleurota, C. rei and *C. gracilis* increased in ground cover over time compared to the non-treated site (0). Caribou or Reindeer lichens (*Cladonia rangiferina* and *C. mitis*) were observed in very small amounts at the zero treatment and young reclaimed sites but were dominate at the 40 years since reclamation (1980) site. The larger Reindeer lichens were out-competing the Pixie-Cup lichens at the oldest reclaimed site. *Stereocaulon* and *Peltigera* spp, both nitrogen fixers, were found at several sites.

Pohlia nutans and *Polytrichum juniperinum* were the major mosses at the younger reclaimed sites and were partially replaced by more woodland mosses, *Dicranum scoparium, Pleurozium schreberi* and other carpet mosses at older sites. The cover of mosses also increased with age of reclamation.

In *Cladonia rangiferina* both copper and nickel concentrations are higher in the relatively barren area with no treatment. Likely soil particles are splashed onto the lichen (Figure 2). Concentrations are lower in all the reclaimed sites with little difference between the age of the site. Background concentrations were found in the forest locality (Figure 2).

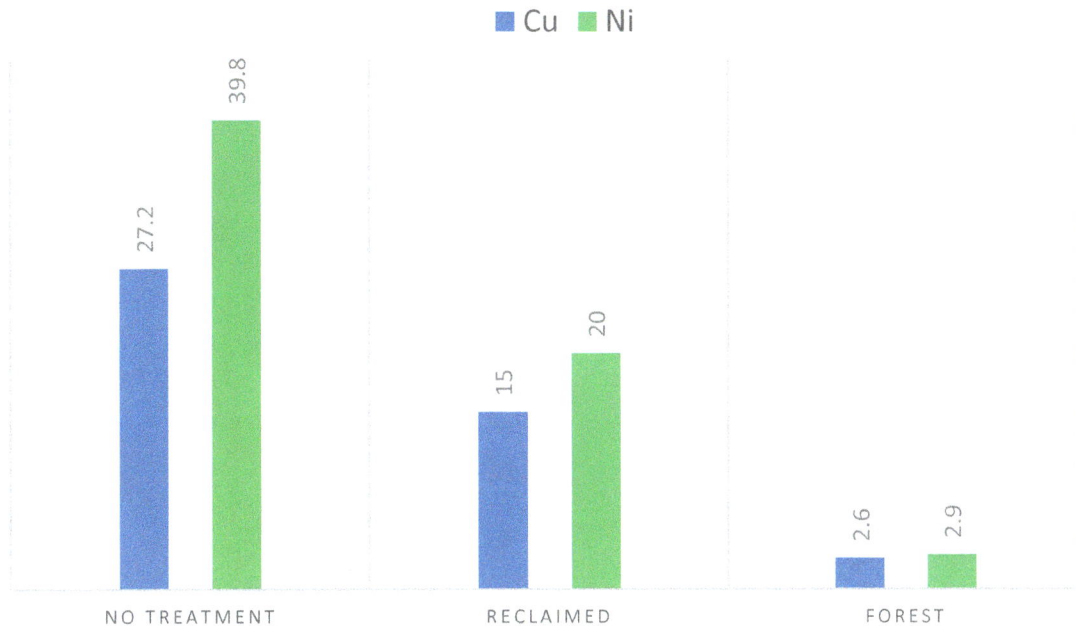

FIG 2 – Copper and Nickel Concentrations (µg/g) in *Cladonia rangiferina* podetia collected from a site with no treatment, five reclaimed sites and a forest outside Sudbury in November 2021.

CONCLUSIONS

The effect of liming, fertilising, grass seeding and tree and shrub planting has provided suitable microsites for the invasion of lichens and mosses. The primary colonisers on both reclaimed and disturbed sites consisted of *Cladonia* or 'Pixie Cup' lichens. With time, with more extensive tree canopy cover the abundance of Cladonia or 'Pixie Cup' species decreases and the *Cladonia* or 'Reindeer' lichens monopolise the area along with a diversity of woodland mosses. For a healthy restored ecosystem, it is important to pay attention to non-vascular species as well as seed-producing plants. This study indicates that mosses and lichen add to the biodiversity of a restored site and should be included in restoration activities. In addition, mosses/lichens could be used as a basis for determining ecosystem functionality and success of revegetation activities.

ACKNOWLEDGEMENTS

We thank Testmark Laboratories in Sudbury for elemental analyses of lichen samples.

Climate change risks to mine closure – planning for a range of impacts

N Bulovic[1], N McIntyre[2] and R Trancoso[3,4]

1. Postdoctoral Research Fellow, The University of Queensland, St Lucia Qld 4072.
   Email: n.bulovic@uq.edu.au
2. Professor of Water Resources, The University of Queensland, St Lucia Qld 4072.
   Email: n.mcintyre@uq.edu.au
3. Science Leader Climate Projections and Services, Department of Environment and Science, Queensland Government, Brisbane Qld 4102.
4. Adjunct Associate Professor of Climate Change, The University of Queensland, St Lucia Qld 4072. Email: r.trancoso@uq.edu.au

INTRODUCTION

Mining is essential for facilitating human development particularly as demand for critical minerals grows. However, over the mine life cycle, the environment is permanently changed. Once mining ceases, mine-affected land can leave significant legacy issues on the surrounding environment if not managed. Therefore, mine closure planning is critical.

Mine closure occurs over large scales often with the aim of recreating resilient ecosystems over mined landscapes with mitigated negative impacts. The natural state of the landscape and the environmental risks over planning time-scales are largely driven by climate. Climate change is expected to impact different parts of the post-mining landscape, from stable landform design (soil erosion mitigation) to revegetation efforts and pit lake management. Consequently, mine closure plans may need to account for a non-stationary and uncertain future climate, that may be both different to the current climate and evolve during the mine life and post-closure.

Here a new data-based approach is presented for incorporating climate change projections in mine closure planning, based on the recently released Coupled Model Intercomparison Project phase 6 (CMIP6) climate data used in the latest Intergovernmental Panel on Climate Change (IPCC) assessment report. The approach: (i) characterises closure-related attributes of the future climate, and (ii) identifies the amount (and direction) of change expected to occur, and its uncertainty, focusing on five metrics for demonstrative purposes and initially describing changes in erosive potential below. This provides information on the type of future environment (ie the changing baseline) that mine closure should plan towards, and identifies rehabilitation options that may be susceptible to increased risks from climate change.

MATERIALS AND METHODS

Overview

The study focuses on expected changes to three climate attributes (and five metrics) important to mine closure and rehabilitation:

1. Extreme rainfall – (i) soil erosion, (ii) floods.
2. Drought – (iii) vegetation establishment, (iv) water availability.
3. Rainfall seasonality – (v) acid mine drainage.

The five listed metrics are derived using downscaled CMIP6 climate projections supplied at a 10 km resolution by the Queensland Future Climate program – an initiative of the Queensland Government to develop application-ready climate data under different emissions scenarios (Chapman *et al*, in review; Syktus *et al*, 2020). Metrics are projected for two future 30-year periods assuming a medium-high emissions scenario (SSP3–7.0): mid-century (2040 to 2069) and end-century (2070 to 2099). Changes in the metrics are estimated relative to a 30-year baseline period (1981 to 2010). The state of Queensland is used as a case study (Figure 1) because of its substantial mining industry, diverse mining regions with varying commodities and climates, and high sensitivity to climate change.

Soil erosion

The initial results presented in Figure 1 focus on one of the five metrics previously listed, namely soil erosion. Climate change impacts soil erosion both directly through rainfall, and indirectly through changes in soil properties, vegetation and land use. Here the direct impacts of climate change on soil erosion are considered through the rainfall erosivity metric (also known as the R-factor). Rainfall erosivity is a commonly used parameter in soil erosion models (such as RUSLE; Renard *et al*, 1997) and is a measure of the mean annual ability of rainfall and run-off to detach and transport a soil particle at a location. It is important to note that rainfall erosivity is not an estimate of soil erosion *per se*, but helps capture the likely direction and magnitude of change due to climate assuming other conditions remain unchanged.

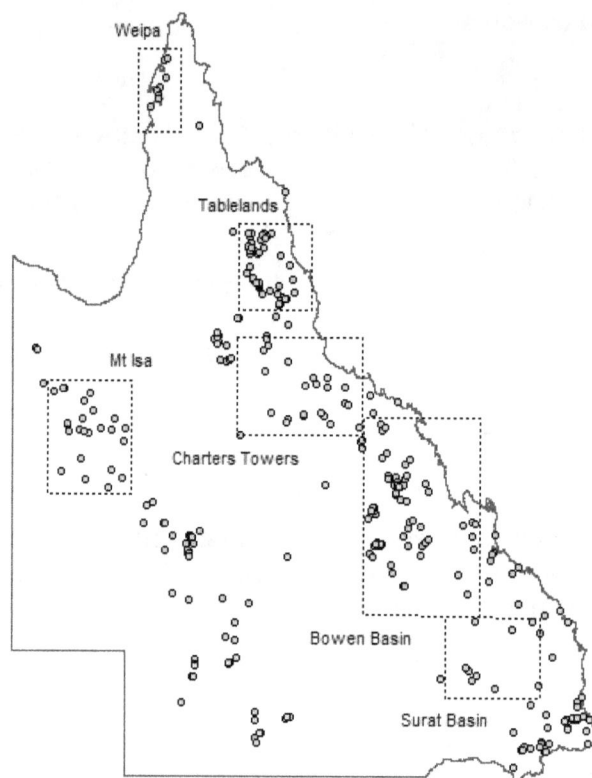

FIG 1 – Map of operating Queensland mines grouped into six key mining regions (Mt Isa, Weipa, Tablelands, Charters Towers, Bowen Basin, Surat Basin). Map based on Worden *et al* (2021).

INITIAL RESULTS

Figure 2 shows boxplots of projected changes in rainfall erosivity across Queensland mine sites at both mid- and end-century. Overall, by mid-century, rainfall erosivity will likely increase across most mine sites (median R = 433), although the magnitude of change varies and is site-dependent (range in R = -188 to 1814). By end-century, increases in rainfall erosivity are projected to occur at three quarters of mine sites; although surprisingly median change is smaller in comparison to the mid-century projections (median R = 221) while the range in values between sites increases because of a couple of outliers (range in R = -1095 to 2508).

Evaluating potential changes in rainfall erosivity by mining region elucidates region-specific characteristics that are not obvious from the Queensland-wide data. For instance, a few notable points are:

- Erosivity will likely increase across mines in the Bowen Basin, Mt Isa, and Surat Basin irrespective of time period.

- Direction of change in erosivity is far more variable for mines in Charters Towers, Tablelands and Weipa because projections indicate positive changes by mid-century, generally becoming negative by end-century.

- Change in erosivity is relatively stable across the Surat Basin.

Importantly, other site attributes need to be considered to translate the changes in rainfall erosivity to soil erosion. For instance, an increase in R of 100 at a sloping site may result in greater soil losses than an increase in R of 1000 over a flat region.

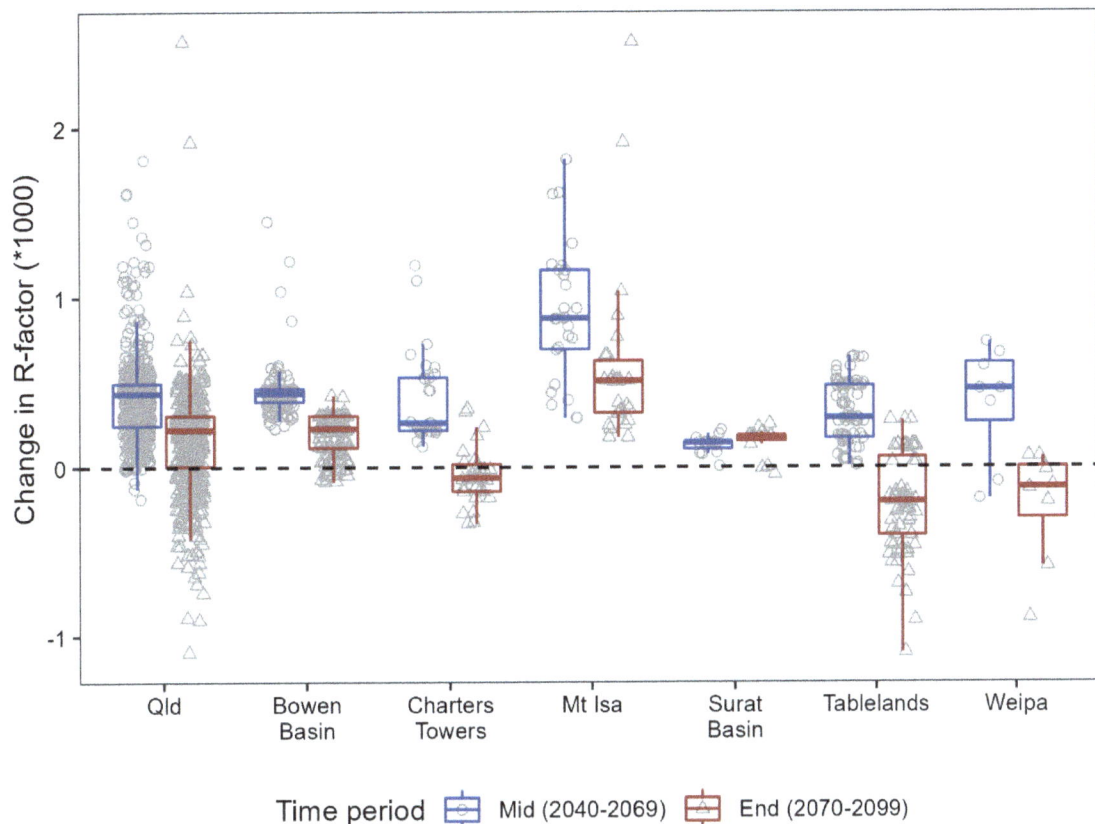

FIG 2 – Boxplots of the range in projected changes in rainfall erosivity at operational mine sites in Queensland, and across individual mine regions, for both the mid- and end-century periods. Points show erosivity values at each individual mine.

FINAL REMARKS

The initial rainfall erosivity results highlight some of the potential complexities when planning for mine closure under climate change, including: the dependence of change direction on location and time-scale; the lack of projections beyond 2099; the need to decide how to handle ensemble (climate model) uncertainty; and the significant differences between CMIP5 and CMIP6 (and possibly future assessments). Evaluation at the level of mining regions is useful in unpacking regional differences. More broadly speaking, combined assessment of the five climate metrics will help highlight locally-relevant risks. Finally, this data-based approach can be expanded to incorporate other important mine closure metrics in collaboration with stakeholders, and be extended spatially beyond Queensland to NSW, Western Australia and other major mining centres to improve knowledge on climate change induced closure risks at a national level.

ACKNOWLEDGEMENTS

The authors acknowledge Dr Pascal Bolz for his help on Queensland mining regions and identifying spatial data sets on mine sites. The Climate Projections and Services team from the Queensland Department of Environment and Science are thanked for their work on downscaling the CMIP6 projections and sharing the data sets in advance of them being made publicly available.

REFERENCES

Chapman, S, Syktus, J, Trancoso, R, Thatcher, M, Toombs, N, Wong, K K-H and Takbash, A, in review. Evaluation of dynamically downscaled CMIP6 models over Australia. EGU General Assembly 2023.

Renard, K G, Foster, G R, Weesies, G A, McCool, D K and Yoder, D C, 1997. Predicting soil erosion by water: a guide to conservation planning with the Revised Universal Soil Loss Equation (RUSLE), Agriculture Handbook No. 703, United States Department of Agriculture, Washington DC.

Syktus, J, Trancoso, R, Ahrens, D, Toombs, N and Wong, K, 2020. Queensland Future Climate Dashboard: Downscaled CMIP5 climate projections for Queensland, accessed from <https://www.longpaddock.qld.gov.au/qld-future-climate/>.

Worden, S, Côte, C, Svobodova, K, Arratia-Solar, A, Everingham, J-A, Asmussen, P, Edraki, M and Erskine, P, 2021. *Baseline works for mine rehabilitation and closure collaboration project*, Sustainable Minerals Institute (The University of Queensland: Brisbane). doi: 10.14264/6c92886

Envisioning the rehabilitation of assets through co-design and gamification

S Emeric[1], E Jones[2], P Baker[3], D Clark[4] and S Grant[5]

1. Global Practice Director – D-Lab, GHD Digital, Melbourne Vic 3000.
   Email: sylvain.emeric@ghd.com
2. Innovation Manager, D-Lab, GHD Digital, Melbourne Vic 3000. Email: emma.jones@ghd.com
3. Executive Advisor – Transactions, Strategy and Commercial, GHD Advisory, Melbourne Vic 3000. Email: phil.baker@ghd.com
4. Australian Mining Leader, GHD, Melbourne Vic 3000. Email: david.clark@ghd.com
5. Manager – Digital Innovation, D-Lab, GHD Digital, Brisbane Qld 4000.
   Email: stephanie.grant@ghd.com

INTRODUCTION

As we strive for a sustainable and resilient future, repurposing assets has become a top priority. In Australia alone, we have a significant number of mineral projects in the pipeline, extensive existing operations, and over 50 000 legacy sites. Collectively, this presents a unique opportunity to be a world leader in mine rehabilitation.

Complex stakeholder landscapes, inflexible policies, aged regulatory frameworks, technical uncertainties and increasing liabilities are all barriers hindering effective transition of legacy assets. The impetus for change is further catalysed by shifting consumer preferences, workforces in flux and regional Australians questioning what the transition means for their local community.

We need innovative approaches to take a leading role in the industry and support global efforts more broadly. Co-designing an ambitious, shared vision spanning projects, operations and legacy assets with all key parties is a fundamental step to unlock and realise maximum societal, environmental, economic and reputational value. Serious play and gamification can further enhance this visioning process.

Co-designing an ambitious, shared vision

Announcements regarding legacy assets can result in shock, denial, anxiety, confusion and anger for employees and communities generationally reliant on mining sites for their jobs and livelihood. Repurposing of these sites also results in positive change, including lowering of pollution and introduction of new technologies that in turn create social value, new skills, and long-term career pathways. To reduce change fatigue and maintain a social license to operate, mining companies need to define their goals and plans for site closure and repurposing in a way that is both clear and consistent. Our experience has shown that a well-defined vision statement is a key ingredient in bringing internal and external stakeholders along complex transition journeys, providing common language to describe the strategic intent behind decisions made for legacy sites and the workforces that support them. Design thinking activities in co-design workshops encourage collaborative and creative alignment on vision, requirements and core components of site transitions. A process of divergent and convergent ideation helps to pose and solve key questions, discuss potential solutions and consider the value that sites can deliver, both now and into the future – all from a human-centric standpoint. The canvassing process translates vision into a set of value propositions that articulate how internal and external stakeholders will benefit from a repurposed site. Once a series of canvases is created, mining companies can phase potential transition initiatives across short-, medium- and long-term horizons according to value and impact.

Gamification to support decision marking

Gamification is a design approach that transforms different systems, services, and activities to resemble games (Koivisto and Hamari, 2019). Following the proliferation and adoption of gaming online and worldwide, research has shown that games help provide cognitive, emotional, social and motivational benefits (Ryan, Rigby and Przybylski, 2006; Granic, Lobel and Engels, 2014). The mining industry is already realising the benefits of gamification, having incorporated elements of simulation and play into existing operations and processes including games that bolster project and asset management, safety education and compliance on-site.

However, more strategic opportunity exists for mining companies to extend their appropriate use of game theory to discuss site transition and closure opportunities and better imagine and anticipate the future of their legacy assets. As the research agenda of bodies like CRC TiME demonstrate, it is time for mining companies to rethink post-mine transitions, evolve their business practices and more deeply empathise with stakeholder perspectives. Testing ideas in the safety of a hypothetical, simulated game environment is crucial given the highly visible, multi-faceted and at times adversarial nature of mine closure announcements and decisions. Through our partnership with Dutch game developer Fresh Forces, we have explored game mechanics that encourage and provoke leaders to think expansively, and hypothetically, in a trusted and psychologically safe environment. By shedding their individual company role for a day, participants can step into the shoes of another internal or external stakeholder and assume a mandate that will impact and influence the future of their asset and site. The game then introduces this mix of mandates to a fresh future, complete with government policies, market conditions, community expectations, local economies, revenue streams and societal priorities.

To facilitate the accuracy and relevance of the game in a mining context, our pilot site transition game is being developed following months of interviews with thought leaders across GHD's global network to reflect a range of perspectives: mining industry, technical, environmental, legal, engineering, client, business, regulatory, governmental, digital, human, creative and cultural. At times, these disciplines directly clash and compete which complicates, confuses, and slows an organisation's decision-making ability. The game recognises and catalyses this, providing a dedicated platform and open forum for different values and views to be advocated for in real time in a fictional environment before decisions are reached. The qualitative interactions between participants of the game carry associated quantitative costs and consequences that impact the financial status and standing of the mining company. Through these interactions, players have discussed options, weighed costs, and debated priorities to reach an agreement on what happens next for the site and asset in focus. After this shared experience, the learnings, insights, and various perspectives on risks and opportunities across different stakeholders can then be re-introduced into real world business. Over the course of one day, the game simulates and expedites content, consultation and collaboration that traditionally take months for organisations to unpack. Whilst a playful process, outcomes are serious and tangible:

- Increase awareness in stakeholders' perspective on risks and opportunities.
- Generate buy-in and social license through trust.
- Boost creative leadership skillsets in empathy, deep listening, negotiation and conflict resolution.
- Save time and cost in feasibility studies by converging faster on the options to be investigated.

Together, gamification and visioning present a fresh and inclusive way to address the complexities and challenges of mine closure and relinquishment. The inherent positivity of these processes and experiences can achieve meaningful results for workforces and organisations in flux and contribute to establishing Australia as a world leader in the mining industry.

ACKNOWLEDGEMENTS

We thank and acknowledge Frisse Blikken and their game studio team as our game development partner.

REFERENCES

Granic, I, Lobel, A and Engels, R C M E, 2014. The Benefits Of Playing Video Games, *Am Psychol*, 69(1):66–78. doi:10.1037/a0034857

Koivisto, J and Hamari, J, 2019. The Rise Of Motivational Information Systems: A Review Of Gamification Research, *International Journal of Information Management*, 45(April 2019):191–210. https://doi.org/10.1016/j.ijinfomgt. 2018.10.013

Ryan, R M, Rigby, C S and Przybylski, A, 2006. The Motivational Pull Of Video Games: A Self-Determination Theory Approach, *Motion and Emotion*, 30(4):347–363. https://doi.org/10.1007/s11031–006–9051–8

In-pit backfilling as part of integrated life-of-mine waste scheduling, closure planning and AMD risk management – Martabe Mine as a case study

S R Pearce[1], K Grohs[2] and A Satriawan[3]

1. Technical Director, Mine Environment Management, Denbigh Wales LL163LU. Email: s.pearce@memconsultants.co.uk
2. Senior Manager – Technical Services, PT Agincourt Resources, Batangtoru, Sumatera Utara 22738. Email: ken.grohs@agincourtresources.com
3. Superintendent – Tailings Projects, PT Agincourt Resources, Batangtoru, Sumatera Utara 22738. Email: azwar.satriawan@agincourtresources.com

INTRODUCTION

Martabe gold and silver mine is operated by PT Agincourt Resources (PTAR) and is situated in northern Sumatra and comprises several deposits in steep terrain, and in a tropical climate, that are required to be developed in the optimal sequence for the best possible Net Present Value (NPV) return on the project. In addition, the waste rock from excavations and tailings from processing is a by-product to be scheduled and used for a number of purposes identified as part of life-of-mine planning. These uses include construction of the current integrated tailings storage facility (TSF) as well as future engineering construction and waste disposal projects which includes plans for a program of in-pit backfilling. These plans need to be aligned to ensure the waste rock and tailings can be contained relative to the production rate.

Given that the site will have multiple open pit voids upon cessation of mining activities opportunities for integration of in-pit disposal of mine waste have now been assessed with technical studies to review the practicality of the approach with respect to operational challenges and conformance with closure planning requirements. This includes assessment of disposal of waste rock, tailings (including consideration of filtered tailings) and co-disposal of waste rock and filtered tailings.

PIT FILLING AND CLOSURE OPTIONS

As part of integrated mine planning, closure planning and waste scheduling and optimisation, PTAR are currently considering three potential scenarios for closure of two currently operating open pits, which includes in-pit waste placement. The three scenarios are as follows:

1. Base case A: pit void left to fill to spill point which forms a pit lake within the void.

2. Option 2B: pit void filled to -20 m of spill point with filtered tailings.

3. Option 2C: pit void filled to -20 m of spill point with filtered tailings and surface placement of co-disposed material up against the pit wall to allow partial backfilling of the pit.

The scenarios are shown in Figure 1 with the extent of backfill shown in plan view and the section view showing the arrangement of the three scenarios in profile.

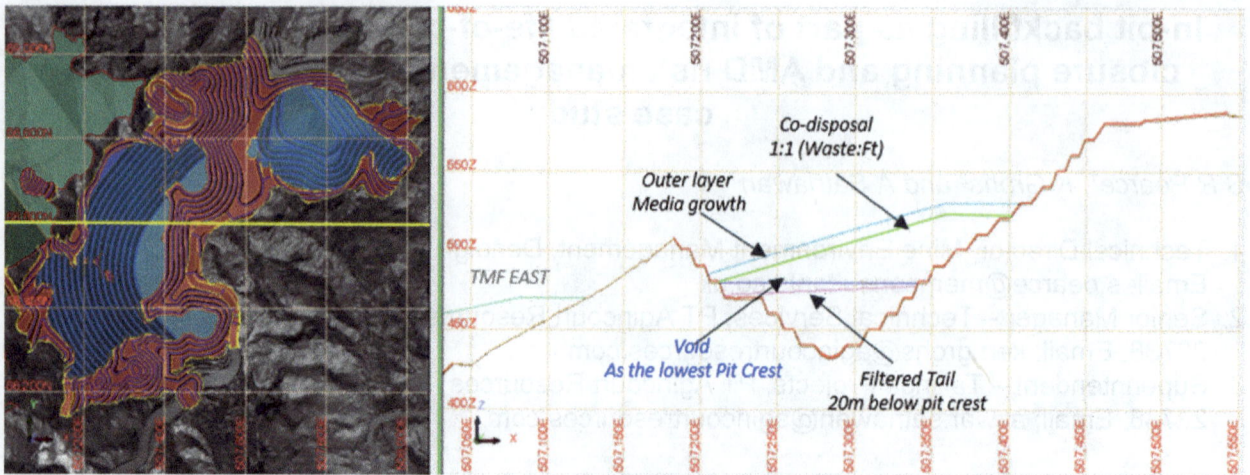

FIG 1 – Waste placement options.

With respect to consideration of closure risks, maintaining surface and groundwater quality leaving site during operation and on-site following closure are key aim for the Martabe Mine. Key risks identified with management of mine waste at the site and with respect to closure of the open pits and potential in-pit filling includes:

- Risk has been identified in open pits that have exposed rock types with higher AMD risk potential in the walls and will generate AMD over the short, medium and long-term.

- Risk has been identified with respect to long-term management of tailings and waste rock at the site which require containment under oxygen limiting conditions due to AMD risks related to sulfide content of the materials.

- Risk identified in mine plan that additional/flexible storage capacity is required in addition to the planned waste storage facilities, for waste rock and tailings deposition to allow optimisation of operational and closure planning.

ASSESSMENT OF AMD RISK FOR IN-PIT CLOSURE AND FILLING OPTIONS

Given that AMD has been identified as a key risk a technical assessment methodology was developed to assess the options identified for pit closure. To allow assessment of AMD risks between the base case option of leaving pit voids open and allowing formation of a pit lake, and options to place waste materials in the pits, a detailed 3D mapping exercise (Figure 2) was carried out utilising existing information held by PTAR operational teams within 3D block models which included:

- Pit shells were analysed within Leapfrog software to derive the potential surface water catchment areas within each pit, the potential discharge locations (for both pit lakes and surface discharges following backfilling) and the expected elevation of these discharge points.

- Projection of the AMD block model onto final pit walls surfaces to allow determination of spatial distribution and surface area of all five classes of waste rock allowing calculation of each waste class exposed on pit walls, both above and below a potential pit lake surface and above an expected backfill level.

- Projection of alteration block model onto final pit wall surfaces to allow determination of spatial distribution and surface area calculation of three alteration types of waste rock both above and below a potential pit lake surface and above an expected backfill level. Alteration type was used to assess the potential risks related to erosion of pit walls and ultimately AMD generation rates.

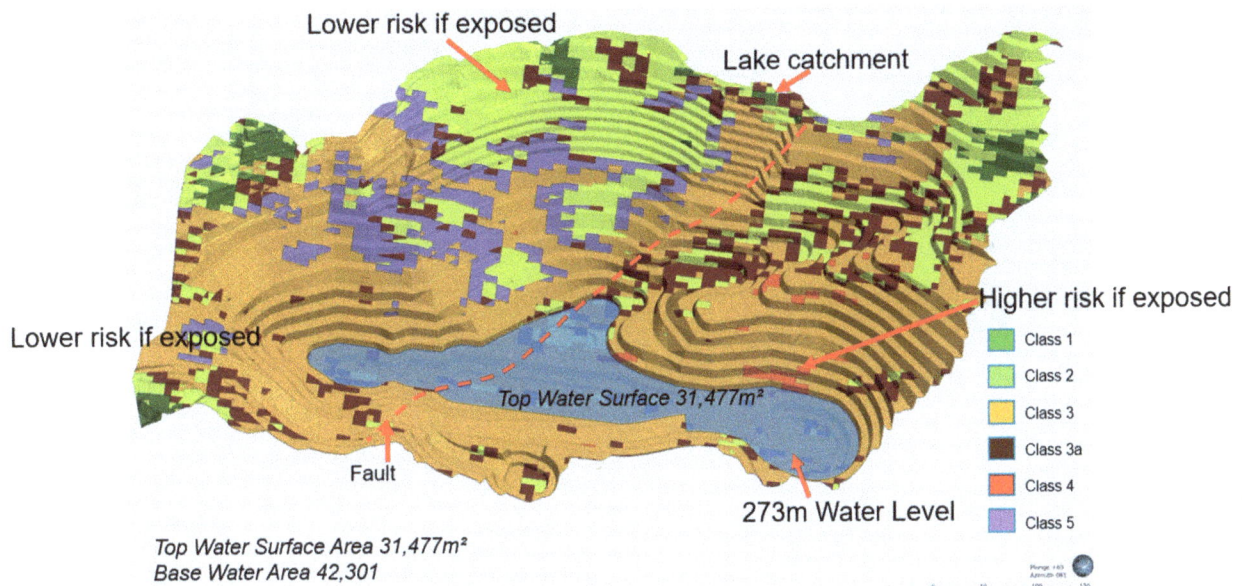

FIG 2 – Base case pit lake, 3D projection of AMD block model onto pit walls.

Under the base case scenario the extent of the pit lake formation is controlled by spill points and the side cut nature of mining, meaning that the majority of pit walls will be exposed into closure. The pit lake area will be a small overall proportion of the overall pit surface area (Figure 2). Given the tropical climate (>4000 mm rainfall per annum) at closure the pit lake would be expected to fill to the discharge level for each pit and then spill to surface water environment.

Under Option 2C the use of mine waste to partially fill the pit void reduces the total area of exposed higher AMD risk material, and eliminates the development of a permanent open water body feature.

The development of a water quality assessment for the various options allows integration of the AMD risk from exposed material with the mass loading produced from this material, and the overall water balance for the system. The results from integrating the water balance with the loading rates and chemistry of non-contact flows water and the quality of the pit surface run-off as well as the final mix with other water flows such as direct precipitation and clean non-contract run-off have been used to assess the various options.

Overall the poorest quality water is produced during the pit lake scenario and the co-disposal scenario suggests the best water quality.

Can biocements stabilise tailings storage facilities?

A E Levett[1,2], S Dressler[3] and G Simpson[4]

1. Geochemist, WSP, Brisbane Qld 4006. Email: alan.levett@wsp.com
2. Honorary Postdoctoral Fellow, The University of Queensland, St Lucia Qld 4072.
   Email: alan.levett@uqconnect.edu.au
3. Senior Landform Designer, WSP, Newcastle NSW 2300. Email: sven.dressler@wsp.com
4. Principal Mine Closure Consultant, WSP, Newcastle NSW 2300.
   Email: gareth.simpson@wsp.com

INTRODUCTION

Tailings storage facilities often represent a mining operation's most significant liability, posing serious environmental, social and safety risks. Almost all mining operations dispose of wet tailings (Valenta *et al*, 2023), which represents a risk of failure due to the potential for a rapid strength reduction of the unconsolidated materials leading to static liquefaction and subsequent dam wall failure. As a waste stream, tailings disposal, decommissioning and relinquishment aim to be inexpensive processes, ruling out many highly engineered solutions. Advantageously, tailings consolidation does not need to be immediate, opening the door for relatively passive, inexpensive consolidation methods that aim to gradually increase the strength of tailings over extended periods (ie years). In this paper, we present the prospect of novel *in situ* tailings stabilisation, learning from studies that have investigated the most erosion-resistant landforms on Earth.

BIOGEOCHEMICAL CYCLING OF IRON

The most erosion-resistant landforms on Earth are the iron-rich surface crusts above banded iron formations (Monteiro, Vasconcelos and Farley, 2018). Micro-organisms drive the physical stabilisation of these environments, constantly cycling iron between relatively soluble ferrous iron (Fe^{2+}) and insoluble ferric iron (Fe^{3+}) in near-neutral pH conditions (Levett *et al*, 2020a). The overall effect of this iron cycling is the consistent formation of new iron-rich cements, which physically stabilise the surficial landforms. Where organic carbon accumulates, micro-organisms promote the reduction of waste iron oxide materials to produce Fe^{2+} (Equation 1).

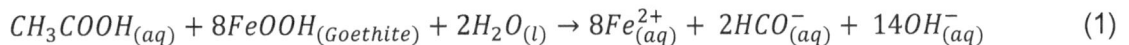

$$CH_3COOH_{(aq)} + 8FeOOH_{(Goethite)} + 2H_2O_{(l)} \rightarrow 8Fe^{2+}_{(aq)} + 2HCO^-_{(aq)} + 14OH^-_{(aq)} \qquad (1)$$

The microbial cell envelope ('cell body') can also provide reactive sites for the re-precipitation of iron-rich minerals. The continued iron oxide precipitation on the cells' surface leads to cellular fossilisation (Figure 1a). The network of micro-organisms (biofilms) naturally grows to bind and aggregate particles. When fossilised, these biofilms form a naturally occurring biocement that is resistant to chemical and physical weathering (Figure 1b). In previous experiments, it has been demonstrated that using a microbial biofilm as an 'organic scaffold' for targeted mineral precipitation can improve the efficiency of cement formation and physical stabilisation by up to 30 times compared with chemical cements that aim to completely infill void space (Levett *et al*, 2020b). For example, less than 1 wt% biocement was required to aggregate iron ore mine waste (Levett *et al*, 2020b), compared with standardised Portland cement, which requires up to 30 wt% cementation materials. These significant increases in cement efficiencies may allow for the *in situ* stabilisation of extensive, unconsolidated tailings storage facilities.

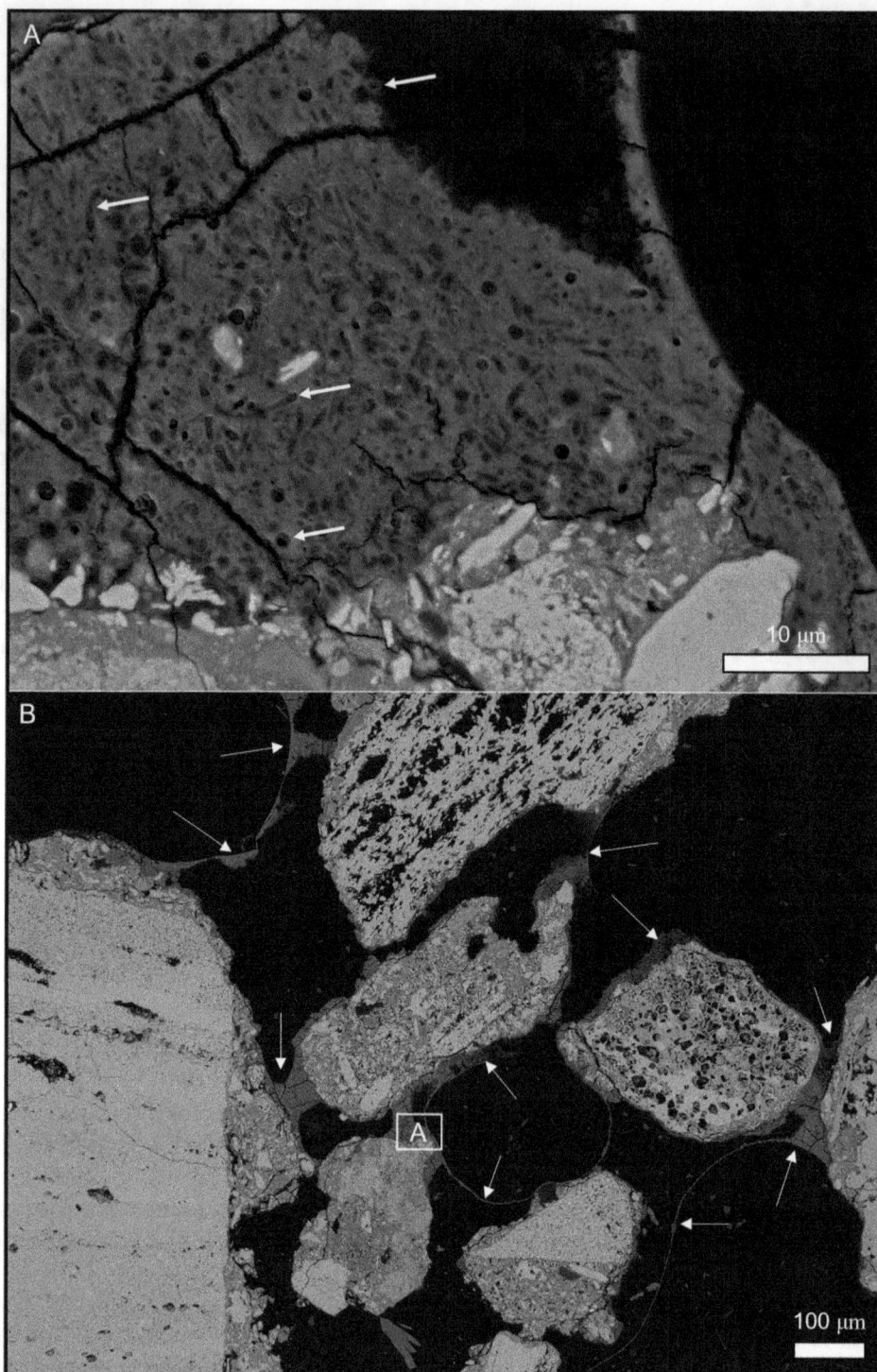

FIG 1 – (a) High magnification backscattered electron (BSE) scanning electron micrograph of microbial cells fossilised by iron oxidation; (b) The low magnification BSE scanning electron micrograph highlights the microbial biofilm that naturally aggregates unconsolidated materials, before being fossilised to create and chemically and physically stable biocement. The white arrows highlight newly formed iron oxide biocements.

STRATEGISING TAILINGS STABILISATION

The major challenge for *in situ* stabilisation of tailings storage facilities is consolidating the subsurface regions, particularly as the hydraulic conductivity of fine-textured tailings can be very low. Consolidation via cementation inherently further reduces the hydraulic conductivity of the tailings; therefore, stabilisation techniques must begin at the lowest required point before progressing towards the surface. Drilling into the subsurface environments of poorly consolidated tailings may introduce an unacceptable level of risk. As such, hydrogeochemical methods that take advantage of the natural seepage are proposed (Figure 2). Initially, Fe^{2+} and degraded carbon from an iron-reducing bioreactor are proposed to be transported by a gravity-fed irrigation system and introduced into the subsurface (>1 m depth) of the tailings storage facility. Introducing the Fe^{2+} into the surface using a gravity-fed system will prevent iron oxidation and precipitation of iron oxide minerals at the surface. A slow-release oxidant may be introduced into the seepage to promote a precisely timed oxidation event, depending on the tailing's hydraulic conductivity and desired depth of mineral precipitation (Figure 2). Examples of environmentally friendly oxidants include ozone (O_3) and potassium ferrate (K_2FeO_4; Sharma, 2002). These oxidants may be coated in a biodegradable polymer, allowing the oxidant to be released after a specific time interval.

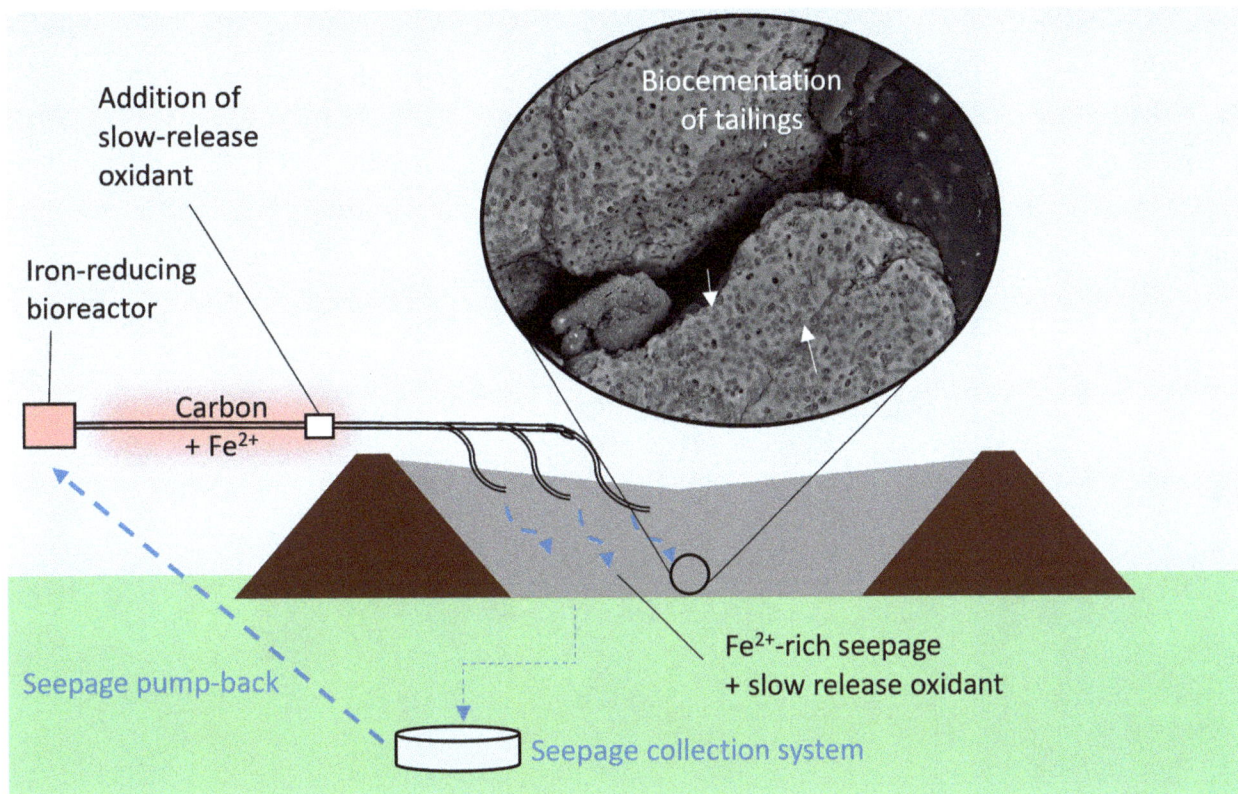

FIG 2 – Conceptual 'turkey nest' tailings storage facility model outlining the process of biocementing surface tailings. Organic carbon, Fe^{2+} and a slow-release oxidant are introduced into the subsurface (>1 m depth) of the tailings to promote: (i) microbial biofilm growth; (ii) iron oxidation; and (iii) microbial fossilisation to stabilise tailings in the subsurface.

CONCLUSIONS

Once the bioreactor solutions are introduced into the subsurface of the tailings, the mechanisms of stabilisation are a three-step process. Initially, the degraded carbon source from the bioreactor promotes the growth of a microbial biofilm, which naturally aggregates the unconsolidated tailings. Secondly, the oxidant is released from the biodegradable polymer, oxidising the Fe^{2+} to Fe^{3+} in the subsurface environment. Lastly, the Fe^{3+} binds to the microbial biofilm promoting the precipitation of iron oxide minerals on the 'organic scaffold' created by micro-organisms, which naturally aggregate materials. Iron oxide cements are chemically and physically stable. Once suitable consolidation of the subsurface is achieved via biocementation, the biodegradable coating's thickness (or material) can be altered to promote iron oxide mineral precipitation closer to the surface. Future work may investigate the resistance of biocemented tailings to liquefaction using large-scale (>1 m) tailings columns. The goal is to improve safety outcomes and allow for successful decommissioning and final relinquishment of tailings landforms.

REFERENCES

Levett, A, Gagen, E J, Vasconcelos, P M, Zhao, Y, Paz, A and Southam, G, 2020a. Biogeochemical cycling of iron: Implications for biocementation and slope stabilisation, *Science of the Total Environment*, 707:136128.

Levett, A, Gagen, E J, Zhao, Y, Vasconcelos, P M and Southam, G, 2020b. Biocement stabilization of an experimental-scale artificial slope and the reformation of iron-rich crusts, in *Proceedings of the National Academy of Sciences*, 117(31):18347–18354.

Monteiro, H S, Vasconcelos, P M P and Farley, K A, 2018. A combined (U-Th)/He and cosmogenic 3He record of landscape armouring by biogeochemical iron cycling, *Journal of Geophysical Research: Earth Surface*, 123(2):298–323.

Sharma, V K, 2002. Potassium ferrate (VI): an environmentally friendly oxidant, *Advances in Environmental Research*, 6(2):143–156.

Valenta, R K, Lèbre, É, Antonio, C, Franks, D M, Jokovic, V, Micklethwaite, S, Parbhakar-Fox, A, Runge, K, Savinova, E, Segura-Salazar, J and Stringer, M, 2023. Decarbonisation to drive dramatic increase in mining waste – options for reduction, *Resources, Conservation and Recycling*, 190:106859.

Microalgae in mining – environmental and social opportunities

A E Levett[1,2] and J Durbin[3]

1. Geochemist, WSP, Brisbane Qld 4006. Email: alan.levett@wsp.com
2. Honorary Postdoctoral Fellow, The University of Queensland, St Lucia Qld 4072.
   Email: alan.levett@uqconnect.edu.au
3. Principal Mine Closure Specialist, WSP, Brisbane Qld 4006. Email: jeremy.durbin@wsp.com

INTRODUCTION

The environmental, social and governance (ESG) obligations of mining companies promote the consideration of water and land resources as well as the post-mining employment opportunities following the cessation of mining operations. This work aims to outline the potential for a mining operation to use their brackish mine water resources to support microalgae production, generating high-quality organic carbon sources to support their mine rehabilitation and closure outcomes. Microalgae production is an established commercial process for high-value products including vitamins, cosmetics and food supplements. On mine sites, microalgae production offers several major environmental and social opportunities: (i) carbon capture, (ii) metalliferous water treatment, (iii) acid mine drainage prevention, (iv) beneficial reuse of brackish mine water, (v) accelerated soil formation and (vi) post-mining employment (Levett et al, 2023).

CARBON CAPTURE

Microalgae's capacity to capture and store atmospheric-derived carbon dioxide (CO_2) may be a key incentive for mining companies, which are likely to rely on carbon offsets to achieve their 'net zero' goals (Levett et al, 2023). Microalgae rapidly sequester CO_2, growing more than ten times faster than terrestrial plants (ie sugarcane) (Schenk et al, 2008). To produce 1 kg of microalgae biomass requires 1.83 kg of CO_2 (Brennan and Owende, 2010) and, in Queensland, microalgae production rates would be expected to exceed 0.015 kg m^{-2} day^{-1} (Roles et al, 2021). Therefore, a microalgae production facility would have the potential to sequester at least 100 t CO_2 ha^{-1} y^{-1}.

BENEFICIAL USE OF SALINE MINE WATER

Microalgae production is likely to be limited to mine sites with sufficient water and non-arable land resources required for the construction of a microalgae growth facility, likely as 'raceway tracks' (Figure 1). In Australia, central Queensland open cut coalmines have been identified as potential early adopters of microalgae production due to their brackish water resources and available land. A simplified conceptual model of an open cut coalmine highlights a potential site layout, using a mine site's near-neutral pH brackish to saline water to produce microalgae in a controlled raceway track growth facility (Figure 1).

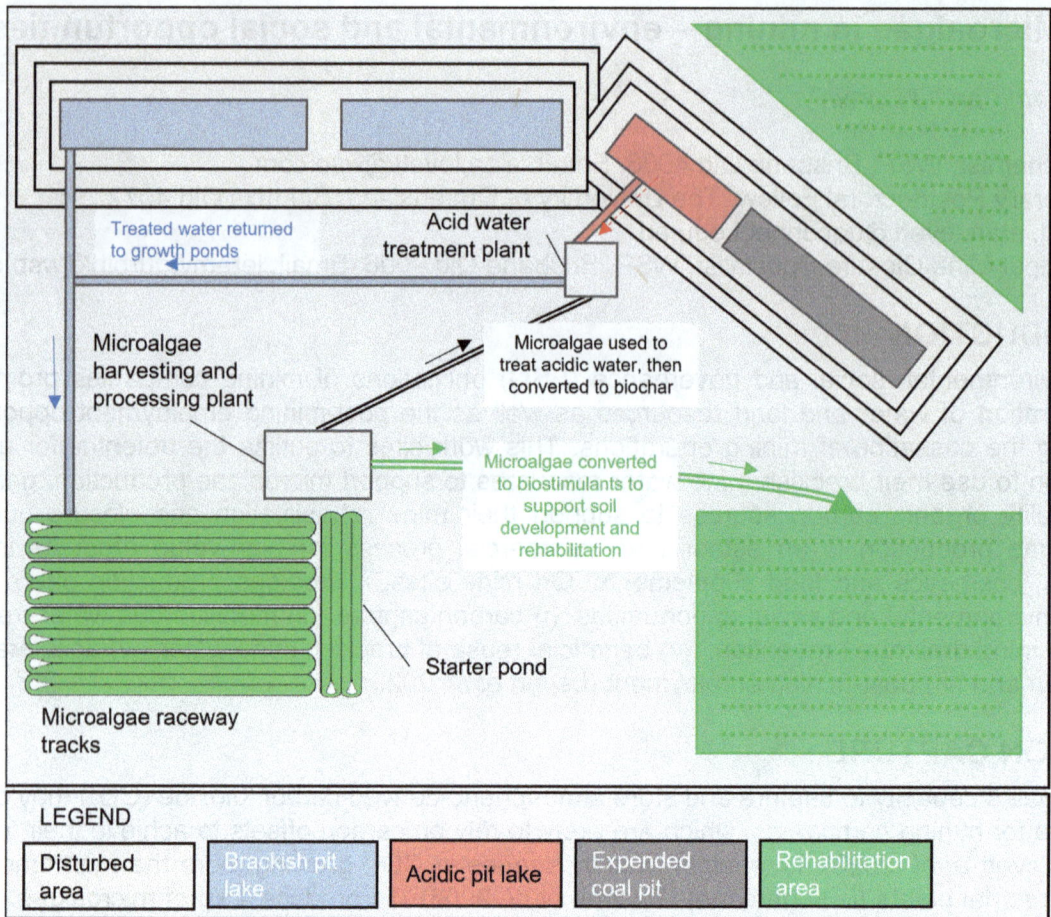

FIG 1 – Simplified conceptual model of an open cut coalmine highlighting the potential to produce microalgae on mine sites using brackish to saline mine water. As well as sequestering CO_2, microalgae can be used to treat acid water and/or used to produce biostimulants to support soil formation and mine rehabilitation.

WATER TREATMENT

One of the most important economic considerations for producing microalgae on mine sites may be the capacity to treat acidic mine waters. Microalgae can treat acidic water via two main mechanisms, referred to as 'passive' or 'active' treatment.

Passive treatment will be applicable for mine sites with large volumes of acidic water. Treatment may be achieved using a dedicated flow-through reactor using dried microalgae (Figure 1). The passive treatment uses dried microalgae to 'passively' bind excess protons (H^+, ie acid) and soluble metals ($Me^{2+/3+}$) in the acidic water to the negative functional groups on the microalgae cells' surface (ie OH^-, NH_2^-, COO^-, PO_4^{3-} etc) via biosorption. Approximately 2 g of dried microalgae is required to treat one litre of acidic mine water (Levett *et al*, 2023). Using a production rate of 0.015 kg m^{-2} day^{-1}, it is expected that 27×10^6 L ha^{-1} y^{-1} of acid water could be treated. For example, a 3-ha microalgae farm would take approximately three years to treat 240×10^6 L of acidic water. After the acidic water has been treated, it can be returned to the microalgae growth ponds and subsequently used to grow microalgae (Figure 1). The microalgae biomass used to treat acidic water can subsequently be converted to biochar to reduce metal mobility (Roberts *et al*, 2017) or co-disposed with other mine waste materials.

Alternatively, active treatment refers to the 'active' bioaccumulation of soluble metals and sulfate into living microalgae (Levett *et al*, 2023). Photosynthetic microalgae can alkalise their growth medium, up to pH ~11 (Equation 1). As such, for mine sites with lower volumes of acidic water, it may be possible to systematically add the acidic water to the microalgae growth ponds to control and optimise the pH between 7.5–9.

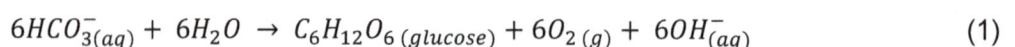

$$6HCO_{3(aq)}^- + 6H_2O \rightarrow C_6H_{12}O_{6\,(glucose)} + 6O_{2\,(g)} + 6OH_{(aq)}^- \tag{1}$$

ACCELERATED SOIL FORMATION

Apart from water treatment options, microalgae production may also contribute to improving mine site rehabilitation outcomes by generating a high-quality soil ameliorant (ie biostimulants), similar to commercially available Seasol®, produced from macroalgae (seaweed). Microalgae biostimulants have been demonstrated to outperform conventional macroalgae biostimulants, improving germination rates by 1.6 times (Rupawalla *et al*, 2022). Microalgae biostimulants do not require dilution for application (Garcia-Gonzalez and Sommerfeld, 2016). Instead, they can be directly combined with waste rocks to a desired carbon and nitrogen proportion to accelerate soil formation processes. Accelerating soil formation using the available waste rock is expected to alleviate soil limitations during mine rehabilitation (Table 1). For many mine sites, this may reduce the amount of material transport and alternate 'borrow' materials required for rehabilitation. In addition, the microalgae soil ameliorants are also expected to increase the soil carbon sequestration and plant biomass on rehabilitation sites (Antonelli *et al*, 2018), potentially providing additional carbon offsets for mining companies.

TABLE 1

The potential financial and environmental, social and governance (ESG) benefits of microalgae production on mine sites.

Mine closure consideration	Microalgae production as part of mine closure	Conventional methods
Acid water treatment	• 'Passive' treatment using dead microalgae in a flow-through plant for mines with large volumes of acidic water (↓cost, ↑ESG) • 'Active' treatment option where acidic water is added directly to microalgae growth ponds to lower pH (↓cost, ↑ESG)	• Acid neutralisation treatment or high-density sludge treatment plant
Rehabilitation	• Convert microalgae to biostimulants to support revegetation (↓cost, ↑ESG) • Reduce the cover volume of materials (↓cost) • Use microalgae as part of an oxygen-consuming cover for potentially acid-forming materials • Accelerate soil formation from mine waste rocks and tailings (↓cost, ↑ESG) • Reduce material transportation requirements for rehabilitation (↓cost) • Increased germination rates (↓cost, ↑ESG) • Convert microalgae used to treat acidic water to biochar (↑cost, ↑ESG)	• 'Soil' materials sourced from borrow locations • Use of thin soil covers and/or blend soil substrates • Poor quality soils with low germination rates • Potentially large transport costs to distribute the desired materials to where they are required
Carbon capture and storage	• Primary carbon capture into the microalgae (~100 t CO_2 ha^{-1} y^{-1}) (↑revenue and ESG) • Accelerated soil carbon sequestration (↑revenue and ESG) • Accelerated plant biomass sequestration (↑revenue and ESG)	• No additional carbon capture.
Post-mining employment opportunities	• Microalgae ponds have a lifespan of ~30 years • 40 ha of microalgae processing would support ~25 employees (↑ESG) • Microalgae potentially used to produce high-value products (cosmetics etc)	• Potential investment into solar, hydrogen/ammonia production etc

Notes: Green highlights indicate potential increases in revenue and ESG outcomes. Light blue highlights indicate potential decreases in site operational costs.

SOCIAL OPPORTUNITIES

Socially, microalgae production facilities have a lifespan of approximately 30 years and are expected to support employment for approximately 25 people, which may continue after the cessation of mining. As such, there are potential economic, environmental and social incentives (Table 1) for a microalgae production facility to support progressive mine rehabilitation during mine operation, after which the facility could be transferred to a third party, providing an alternative industry in regional areas for the balance of the microalgae facilities lifespan.

REFERENCES

Antonelli, P M, Fraser, L H, Gardner, W C, Broersma, K, Karakatsoulis, J and Phillips, M E, 2018. Long-term carbon sequestration potential of biosolids-amended copper and molybdenum mine tailings following mine site reclamation, *Ecological Engineering*, 117:38–49.

Brennan, L and Owende, P, 2010. Biofuels from microalgae—a review of technologies for production, processing and extractions of biofuels and co-products, *Renewable and sustainable energy reviews*, 14(2):557–577.

Garcia-Gonzalez, J and Sommerfeld, M, 2016. Biofertilizer and biostimulant properties of the microalga *Acutodesmus dimorphus*, *Journal of applied phycology*, 28:1051–1061.

Levett, A, Gagen, E J, Levett, I and Erskine, P D, 2023. Integrating microalgae production into mine closure plans, *Journal of Environmental Management*, 337:117736.

Roberts, D A, Cole, A J, Whelan, A, de Nys, R and Paul, N A, 2017. Slow pyrolysis enhances the recovery and reuse of phosphorus and reduces metal leaching from biosolids, *Waste Management*, 64:133–139.

Roles, J, Yarnold, J, Hussey, K and Hankamer, B, 2021. Techno-economic evaluation of microalgae high-density liquid fuel production at 12 international locations, *Biotechnology for Biofuels*, 14(1):1–19.

Rupawalla, Z, Shaw, L, Ross, I L, Schmidt, S, Hankamer, B and Wolf, J, 2022. Germination screen for microalgae-generated plant growth biostimulants, *Algal Research*, 66:102784.

Schenk, P M, Thomas-Hall, S R, Stephens, E, Marx, U C, Mussgnug, J H, Posten, C, Kruse, O and Hankamer, B, 2008. Second generation biofuels: high-efficiency microalgae for biodiesel production, *Bioenergy research*, 1:20–43.

The critical pool levels for the Yallourn Mine determined using the MGRI approach

S Narendranathan[1], N Patel[2], J Butler[3] and S Rastogi[4]

1. MAusIMM(CP), Senior Technical Director (Mining), GHD, Traralgon Vic 3844.
   Email: sanjive.narendranathan@ghd.com
2. MAusIMM, Senior Geotechnical Engineer, GHD, Melbourne Vic 3000.
   Email: nirav.patel@ghd.com
3. Geotechnical Engineer, GHD, Melbourne Vic 3000. Email: james.butler@ghd.com
4. Geotechnical and Geology Leader, Energy Australia Yallourn Mine, Yallourn Vic 3825.
   Email: sid.rastogi@energyaustralia.com.au

INTRODUCTION

The Yallourn Mine (YM) commenced mining of brown coal in the 1920s under operation by the State Electricity Commission of Victoria (SECV), prior to it becoming privatised in 1996. The present owners of the YM, Energy Australia Yallourn (EAY), acquired the site in 2000 and intend to continue operations until scheduled closure in 2028. Brown coal extracted from the YM is fed into the Yallourn Power Station which supplies approximately 22 per cent of Victoria's electricity.

Water within the YM is collected and stored in the Township Field *Fire Service Pond*. Water levels within the Fire Service Pond (FSP) may fluctuate considerably depending on the requirement for storage at the site.

With due consideration to the stability performance of mine batters adjoining the FSP, the authors performed slope stability assessments to evaluate the risk and consequence of batter instability during the potential pond filling as it approaches the 'critical pool'. The critical pond is defined as the level at which the driving forces (destabilising forces) are at their greatest and the resisting forces (lake forces, friction) are at their lowest. This assessment was undertaken using the Mine Geotechnical Risk Index (MGRI) approach.

KEY STABILITY CONSIDERATIONS

Mining of the brown coal along the Fire Service Pond (FSP) Batter Domain presents several operational challenges with respect to stability which require specific consideration:

- The geological conditions at the site (eg tectonic activity, depositional geometry, variability in the mechanical characteristics of the coal seams and associated interseams and the presence of defects).

- Historic mine batter stability performance, including past episodes of ground movements resulting in notable variability of material strength characteristics within the coal and interseam units.

- Historic remediation of coal oxidation/smoulder event resulting in deterioration of coal strength, referred to as 'shattered coal', in localised areas.

- Fluctuations in FSP water level inducing varying phreatic conditions throughout the FSP mine batters.

Where the above factors are not suitably evaluated the potential for unplanned batter movement resulting in disruption to mine operations may occur. The likelihood of unplanned batter movement is exacerbated at the critical pool level and so particular caution must be applied during these periods.

Mechanics of failure

The critical instability mechanism which can impact overall slope scale at the YM is 'coal block sliding' (Figure 1). This mechanism can result where there are excessive destabilising forces such as the below:

- Hydrostatic pressures within the coal joints/tension cracks.

- Elevated phreatic gradient within the coal unit.

- Upthrust from underlying aquifers (floor heave).

- Localised zones of shattered coal ie reduced strength coal, due to historic coal oxidation and subsequent remediation. It should be noted that within the shattered coal, localised instability mechanisms other than coal block sliding may occur.

The above destabilising forces are typically counteracted by the following stabilising forces:

- The strength of the underlying interseam, the unit upon which the overlying coal block could 'slide'.

- Resisting forces afforded by pit lake forces (ie acting on the batter).

- Floor heave due to insufficient weight balance (ie for lower lake level or empty void) which can occur where unmanaged aquifer upthrust pressures are greater than overlying water weight.

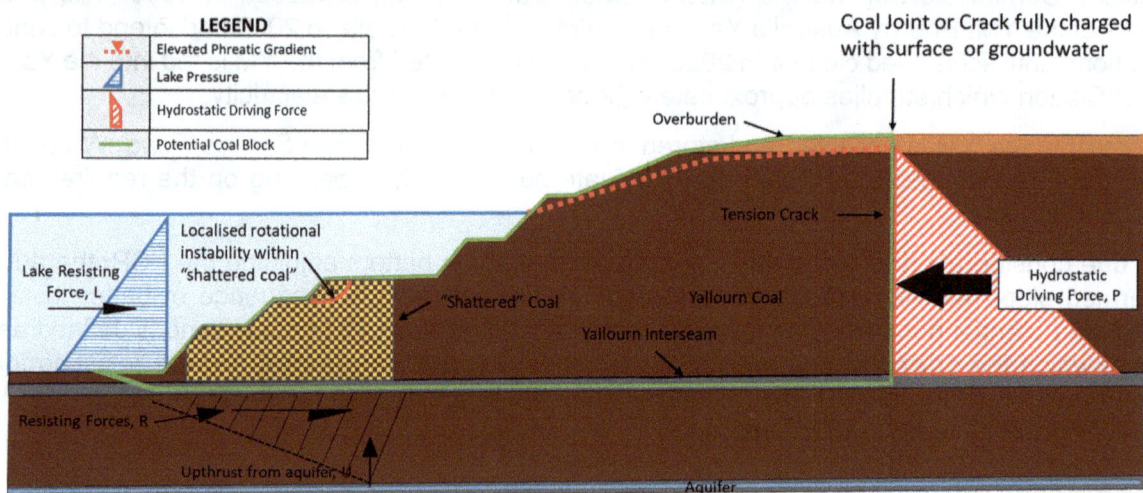

FIG 1 – Coal block sliding mechanism.

Definition of problem

With respect to the complexities surrounding the FSP Batter Domain, including past ground movement episodes and presence of shattered coal in localised zones, it can be appreciated that there may exist one or multiple pond levels which fail to sufficiently counteract hydrostatic (horizontal) and aquifer uplift forces ie destabilising forces. This is defined by the authors as the critical pool(s).

Failing to appreciate the critical pool and the associated stability implications can pose significant challenges to the overall objective of achieving satisfactory stability during pond filling. This has previously been reported in the study undertaken by Narendranathan *et al* (2022). Furthermore, it provides insights to slope design practitioners in relation to implementing suitable amelioration measures in the short to medium term.

Analytical methodology

The authors have analytically determined the occurrence of a critical pool utilising probabilistic 2D limit-equilibrium modelling. Input data included groundwater levels obtained through monitoring and mechanical properties of the geological units interpreted from laboratory test results. The below factors were calculated:

- Factor of Safety (FoS) – a ratio of the resisting forces against the destabilising forces. A FoS of less than 1.0 indicates that the sum of the driving forces is greater than the sum of the resisting forces.

- Probability of Failure (PoF) – The probability of failure is defined as the likelihood of the FoS falling below 1.0.

A key limitation in the above conventional stability metrics is the inability to objectively attribute a consequence and seasonal (temporal) overlay to the resulting (FoS, PoF) outcomes. Narendranathan and Cheng (2019) developed the MGRI to evaluate transitional stability by calculating 'windows' of risk increases and comparing the outcomes against tolerable risk thresholds (Narendranathan *et al*, 2022). Where the ensuing stability calculations exceed the tolerance criteria, it is identified as 'critical,' and hence enables the identification of range of critical pools.

Slope stability analyses was undertaken on two areas from the FSP Batters Domain, representative of batters in which competent coal (FSP02) and shattered coal (EW4) are encountered. Depicted below in Figure 2 are the resultant calculations employing the MGRI methodology. The results indicate:

- For FSP02, the critical pool was identified to occur between RL +3 m and +10 m Australian Height Datum (AHD).

- For EW4, where shattered coal is encountered, the critical pool was identified to occur between RL -3 m and +3 m AHD.

This duality in the critical pool concept is particularly unique to this area of the YM, and understanding the consequence associated with the stability outcomes when the mine is operating at this range enables the development of suitable management protocols when considering the temporal exposure at these pond levels.

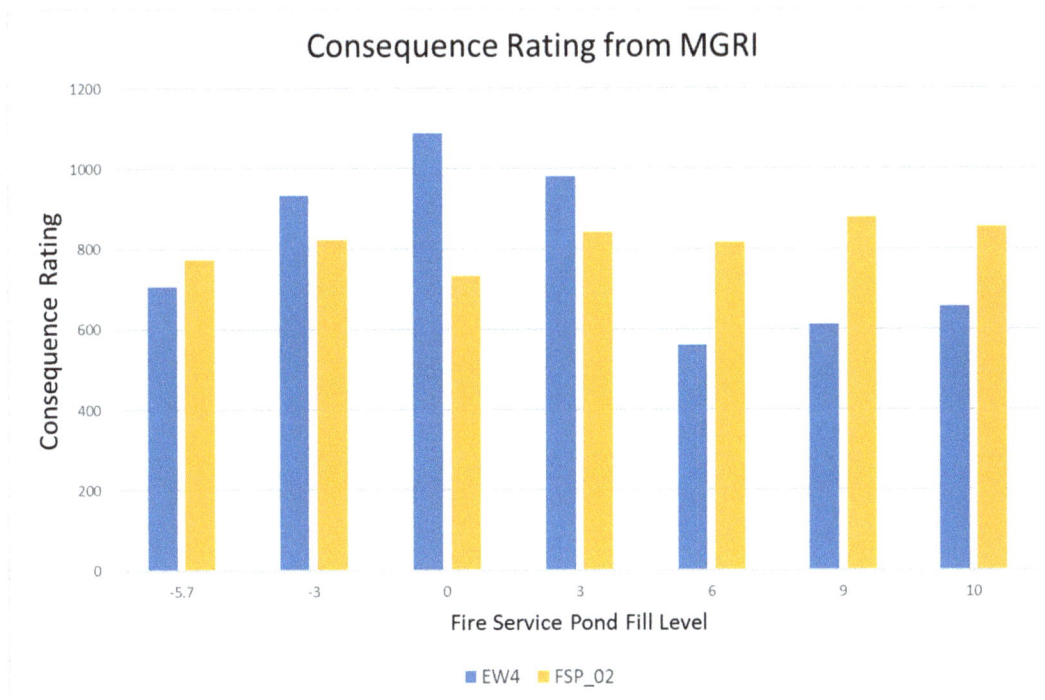

FIG 2 – Consequence ratings across YM from MGRI.

CONCLUSIONS

Employing the MGRI process the authors were able to identify the presence of two critical pools and accordingly suitably manage the stability of the surrounding FSP batters in the vicinity of these pond levels. The authors note that at other mines in the Latrobe Valley, the critical pool level typically occurs over a smaller range, and tend to be unique values. for a given interseam, however, at the YM, and particularly along the FSP Batters Domain, the critical pool occurs in two distinctly dissimilar points. This is due to the unique attributes associated with the FSP Batter Domain as outlined previously. As noted by Narendranathan and Cheng (2019) in the development of the MGRI, the MGRI can be applied when consequence of failure may be highly variable and risk level difficult to define to provide an objective approach to rationalise geotechnical management.

ACKNOWLEDGEMENTS

The authors would like to thank EnergyAustralia Yallourn, in particular Lance Wallace, and GHD for providing the opportunity to publish this paper.

REFERENCES

Narendranathan, S and Cheng, M, 2019. Development of the Mine Geotechnical Risk Index, in *Proceedings of the First International Conference on Mining Geomechanical Risk* (ed: J Wesseloo), pp 461–474 (Australian Centre for Geomechanics: Perth).

Narendranathan, S, Patel, N, Mok, T, Stipcevich, J and Symonds, A, 2022. A case study: conceptual rehabilitation and closure planning for the Loy Yang mine using the risk-based probabilistic approach employing the Mine Geotechnical Risk Index methodology, in *Mine Closure 2022: 15th International Conference on Mine Closure* (eds: A B Fourie, M Tibbett and G Boggs), pp 449–462 (Australian Centre for Geomechanics: Perth).

The critical pool levels for the Hazelwood Mine determined using the MGRI approach

S Narendranathan[1], N Patel[2], T Mok[3] and J Lowe[4]

1. MAusIMM (CP), Senior Technical Director (Mining), GHD, Traralgon Vic 3844.
   Email: sanjive.narendranathan@ghd.com
2. MAusIMM, Senior Geotechnical Engineer, GHD, Melbourne Vic 3000.
   Email: nirav.patel@ghd.com
3. Geotechnical Engineer, GHD, Melbourne Vic 3000. Email: tanya.mok@ghd.com
4. Head of Regulation, Compliance and Sustainability, Engie, Melbourne Vic 3000.
   Email: jamie.lowe@engie.com

INTRODUCTION

The Hazelwood Mine (HM) commenced operations in the 1950s and was originally operated by the State Electricity Commission of Victoria (SECV) prior to it becoming privatised. ENGIE, the current owners of the Hazelwood Mine ceased operations in 2017 and have since transitioned the site into the rehabilitation and closure phase, which is to encompass a full pit lake as part of the final landform.

A key aspect of HM closure is the maintenance of long-term stability of the pit slopes to deliver a final landform which is safe and sustainable, particularly given its proximity to critical public receptors. The Princes Freeway, which is the primary access route to the Latrobe Valley (LV), and the Morwell Township are both located immediately to the north of the HM.

With due consideration of the mine void and nearby public receptors, the authors performed slope stability assessments of critical mine domains to evaluate the risk and consequence of batter instability during the transient phase.

TECHNICAL BACKGROUND

Mechanics of failure

The critical instability mechanism which can impact overall slope scale at HM is 'coal block sliding' (Figure 1). This mechanism can result where there are excessive destabilising forces such as:

- Hydrostatic pressures within the coal joints/tension cracks.
- Elevated phreatic gradient within the coal unit.
- Upthrust from underlying aquifers (ie floor heave).

The above destabilising forces are typically counteracted by the following stabilising forces:

- The strength of the underlying interseam, the unit upon which the overlying coal block could 'slide'.
- Resisting forces afforded by pit lake forces (ie acting on the batter).
- Floor heave due to insufficient weight balance (ie for lower lake level or empty void) which can occur where unmanaged aquifer upthrust pressures are greater than overlying water weight.

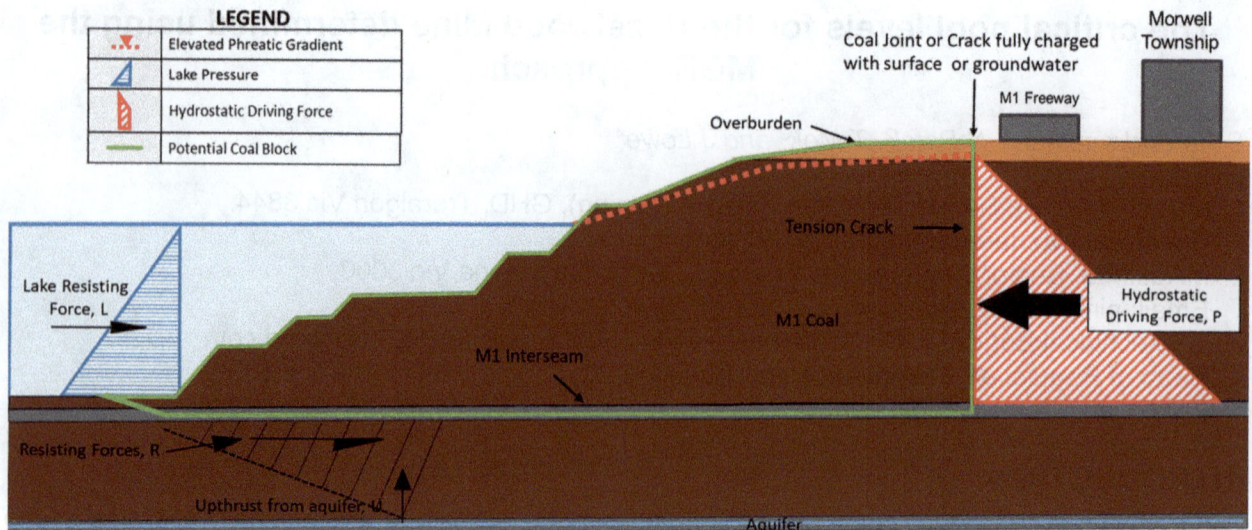

FIG 1 – Coal block sliding mechanism.

This mechanism can result when the destabilising forces (hydrostatic driving force) overcome the resisting forces (lake force, floor heave). The destabilising forces can be exacerbated due to poor management of groundwater and surface water which can cause elevated phreatic conditions within the batters.

Critical pool

As part of the transient phase of mine closure, the authors have identified the lake levels at which the driving forces (destabilising forces) are at their greatest and the resisting forces (lake forces, interseam shear strength) are at their lowest. This is deemed the critical pool level. Failing to appreciate the critical pool and the associated stability implications can pose significant challenges to the overall objective of achieving a stable landform. Additionally it provides insights to slope design practitioners in relation to implementing suitable amelioration measures, to improve slope stability conditions.

Assessing the batter stability risks was undertaken using the Mine Geotechnical Risk Index (MGRI) approach, which is detailed below.

KEY STABILITY CONSIDERATIONS

Mining of the brown coal at the HM presents a number of stability related challenges which require specific consideration:

- The geological conditions at the site – tectonic activity, depositional geometry, variability in the mechanical characteristics of the coal seams and associated interseams, presence of defects.

- Groundwater and surface water conditions.

- History of mining, mining practices, changes in stress conditions.

The HM has been primarily mined through three key geological units including overburden, a coal seam (up to 80 m thick in places) and the associated underlying interseam. Where the above factors are not suitably designed for or managed, they can present adverse geotechnical risks, which are exacerbated at the critical pool level.

METHODS

The authors have analytically determined the occurrence of a critical pool utilising probabilistic 2D limit-equilibrium modelling. The below conventional factors were calculated.

- Factor of Safety (FoS) – a ratio of the resisting forces (interseam shear strength and lake force) against the destabilising forces (hydrostatic pressure within coal joints, elevated phreatic conditions, aquifer uplift), whereby a FoS of less than 1.0 indicates that the sum of the driving forces is greater than the sum of the resisting forces.

- Probability of Failure (PoF) – The probability of failure is defined as the likelihood of the FoS falling below 1.0.

In this paper, the authors have then utilised the MGRI (Narendranathan *et al*, 2021) to objectively attribute a consequence and seasonal (temporal) overlay to the FoS and PoF outcomes from the 2D modelling. The MGRI is defined as:

$$MGRI = (PC_t \times C_f \times I_f \times S_f)$$

Where:

MGRI	= Mine Geotechnical Risk Index
C_f	= Consequence Factor – Probability of Failure (PoF) × Volume of Failure (after Lilly, 2000; Narendranathan, 2009)
PC_t	= Annual probability of the primary instability load or scenario occurring
I_f	= Impact factor
S_f	= Seasonal factor

Narendranathan and Cheng (2019) developed the MGRI to evaluate slope stability by calculating 'windows' of risk increases and comparing these outcomes against tolerable risk thresholds, whereas the FoS and PoF calculations solely do not provide this opportunity. Where the resultant outcomes exceed the tolerance criteria, it is identified as 'critical', and hence enables the identification of range of critical pools. This also enables the optimisation and installation of suitable slope supplementation measures to improve slope stability performance, to meet the design acceptance criteria in line with acceptable risk thresholds.

All mine domains at HM were subjected to detailed 2D slope stability analyses to then assess the MGRI outcomes as a result of a filling mine void. Depicted below in Figure 2 are the resultant calculations from the MGRI. The outcomes of the assessment indicate:

- For the northern batters of HM, the critical pool has been successfully identified by the authors to occur between +5 and +11 m RL.

- This indicates that due to the elevated slope stability risks at and near the critical pool level, robust and increased geotechnical monitoring (pin survey, groundwater levels) and management (drainage holes) is required.

- Measures (eg increased rate of fill) can be implemented to minimise the temporal 'risk' exposure at or near the critical pool level, thereby reducing the duration spent at this juncture.

- The critical pool concept is particularly unique to the LV mines as mine slopes hosted within hard rock are typically less sensitive to the weight differences in comparison to the three LV mines due to the relatively low coal density (ie similar to that of water) and lake counterforces. Accordingly, this requires informed and careful assessment of slope stability during the transient phase.

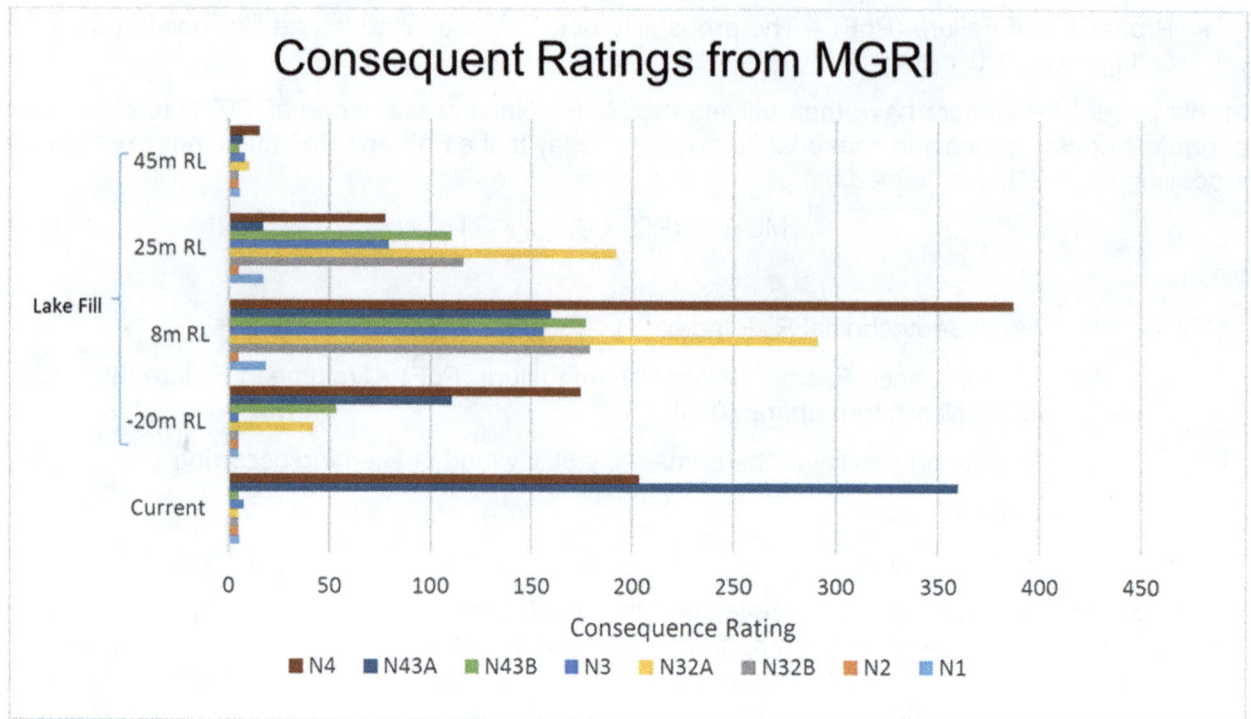

FIG 2 – Consequence ratings across HM from MGRI.

CONCLUSIONS

The authors note that due to the variability in geological conditions and geometrical considerations, the critical pool level can vary between adjacent mine domains and as such multiple critical pool levels occur across the HM. These critical pools can compromise stability conditions without the installation of the suitable slope supplementation measures.

The authors adopted a risk-based approach, the MGRI, developed by Narendranathan and Cheng (2019) to assess the critical pool level, the associated geotechnical significance and extent so as to optimise placement of slope supplementation measures to facilitate the safe filling of the pit up to and beyond these points, as they would otherwise present unacceptable increase in risk.

ACKNOWLEDGEMENTS

The authors would like to thank ENGIE and GHD for providing the opportunity to publish this paper.

REFERENCES

Lilly, P A, 2000. The minimum total cost approach to optimise pit slope design, Western Australia School of Mines, Kalgoorlie.

Narendranathan, S and Cheng, M, 2019. Development of the Mine Geotechnical Risk Index, in *Proceedings of the First International Conference on Mining Geomechanical Risk* (ed: J Wesseloo), pp 461–474 (Australian Centre for Geomechanics: Perth).

Narendranathan, S, 2009. Fundamentals of probabilistic slope design and its use in pit optimization, Proceedings of the 43rd US Rock Mechanics Symposium & 4th US - Canada Rock Mechanics Symposium, American Rock Mechanics Association, Alexandria.

Narendranathan, S, Faithful, J, Patel, N and Stipcevich, J, 2021. A case study – from operation to closure, transient slope supplementation measures for the northern batters of the Hazelwood Mine, using the MGRI approach, in *Mine Closure 2021: Proceedings of the 14th International Conference on Mine Closure* (eds: A B Fourie, M Tibbett and A Sharkuu), QMC Group, Ulaanbaatar. https://doi.org/10.36487/ACG_repo/2152_54

Synergies between renewable-powered mines and community development programs throughout mine life cycle and post-closure

L Rollin[1], J Joughin[2] and D Kyan[3]

1. Senior Consultant, SRK Consulting (Australasia) Pty Ltd, Perth WA 6005.
   Email: lrollin@srk.com.au
2. Corporate Consultant, SRK Consulting (UK) Ltd, Cardiff CF10 2HH, Wales.
   Email: jjoughin@srk.co.uk
3. Principal Consultant, SRK Consulting (Australasia) Pty Ltd, Brisbane Qld 4000.
   Email: dkyan@srk.com.au

INTRODUCTION

In line with the 2015 Paris Agreement and corresponding climate disclosure requirements, many mining companies have publicly disclosed that they are developing and/or implementing ambitious decarbonisation strategies. Sometimes these strategies can be linked to the development of community programs and/or the creation of a positive legacy post-closure.

Renewable power is key to most decarbonisation strategies – some mines will get this from external sources and others will establish on-site power generation and storage facilities. A range of options are available to develop renewable power generation (eg solar panel, wind farm, hydropower) and energy storage (eg battery, hydrogen, pumped hydroelectricity). Land access, impact assessments, planning and environmental approvals are required to establish new facilities. Host communities are more likely to be in favour of, or at least will accommodate, new power supply developments if these are planned with a benefit to the communities.

Transitioning to renewables raises new challenges. Among these are high upfront costs and access to capital, energy demand variability, supply intermittency and variability, storage limitations, grid stability and reliability, transmission and distribution infrastructure constraints. Compounding these challenges are issues such as infrastructure resilience to extreme weather events (eg heat, wind), land use and access, lengthy permitting processes, community perceptions, skilled labour requirements, and meeting mine closure objectives.

Renewable energy sources provide a more decentralised source of energy that reduces reliance on imported fuels, thereby increasing energy security and lowering production costs. On the contrary, implementing renewable energy could lead to potential negative impacts such as visual perception from the community, competitive land usage, electricity price raises due to upgrading infrastructure and the use of electricity storage capacities.

As part of the definition of an environmental and social management plan, participative stakeholder engagement is required to identify benefits to the communities. For example, among the recognised benefits of developing solar panels (increase job opportunities, reduce soil erosion), various technologies have been experimented globally for the last ten years and are emerging due to their proven community benefits. Key findings to enhance those benefits can range from the development of innovative practices such as 'agri-/aqua-voltaic', 'solar grazing' to optimise land usage and stimulate the local economy (Table 1), to the implementation of technologies providing services to communities such as mobile water desalination plants or 'smart grid' systems to minimise energy clipping during peak-load of oversized renewable energy facilities (Table 2).

While developing those synergies, careful management is required to avoid creating superfluous energy service needs from community programs. Energy sufficiency should remain a key principle according to the sufficiency, efficiency and renewable framework of the Intergovernmental Panel on Climate Change. Darby and Fawcett (2018) define energy sufficiency has '_a state in which people's basic needs for energy services are met equitably and ecological limits are respected_'. Sufficiency tackles the causes of the environmental impacts of human activities by avoiding the demand for energy and materials over the life cycle of buildings and goods. Efficiency tackles the symptoms of the environmental impacts of human activities by improving energy and material intensities. The third pillar, renewables, tackles the consequences of the environmental impacts of human activities by reducing the carbon intensity of energy supply (Cabeza _et al_, 2022).

This paper provides project developers, decision-makers and regulatory authorities with ideas on synergies between renewable energy options and community development programs. These can be considered to enhance stakeholders' support for implementation of decarbonisation strategies.

TABLE 1

Extract of solar photovoltaic electricity generation technologies and associated benefits.

Innovative practice	Benefit examples	To project	To community
Solar grazing Dual use of grassland for solar farming and grazing underneath solar panels	• Control vegetation: alternative to lawnmowers, livestock can reach around legs of structures	X	
	• Reduce wildfire risk by controlling vegetation around solar panels, a concern in arid regions	X	
	• Enhance or stimulate local economic growth: lower herbage mass available in solar pastures is offset by higher forage quality, can increase local meat or dairy production		X
Agrivoltaic, agrisolar, agrophotovoltaics Dual use of crop areas for solar farming (panels vertical or placed on roof of greenhouse) and agriculture	• Enhance or stimulate local economic growth: shade can reduce heat stress and water loss and can increase crop yields and allow crops to grow during dry seasons		X
	• Increase water efficiency: shade increases freshness and reduces evaporation and water requirements for crops		X
Aquavoltaic Dual use of water body for solar farming (floatovoltaic) and aquaculture	• Improve power conversion efficiency due to water cooling and cleaning of solar panels	X	
	• Enhance or stimulate local economic growth: solar panel systems can provide filtered light to maintain water temperatures and improve fish growth rate		X
	• Increase water efficiency: shade reduces the evaporation rate of ponds		X
Common benefit to those practices	• Reduce land use pressure: land used for dual production of food and electricity reduces the need for land conversion	X	X
	• Educational opportunities to provide information about renewable energy, sustainability and environmental stewardship	X	X

TABLE 2

Extract of technologies which can provide community services.

Technology	Benefit examples	To project	To community
Mobile desalination plant and storage Enhance the use of peak-load energy to provide fresh water to isolated regions	• Increase access to clean water: convert sea water or brackish water to drinking water for isolated communities		X
	• Increase food production: desalinated water can be used for irrigation, a concern in arid regions		X
	• Reduce reliance on surface water sources and water transportation		X
Smart grid and vehicle-to-grid system Enhance the use of peak-load energy by using electric vehicles' battery storage	• Improve grid resilience and flexibility: the batteries of electric vehicles can function as a source of backup power during power outages or grid disturbances and fluctuations, reducing the likelihood of blackouts, improving overall grid stability, reliability and cost-effectiveness of the grid	X	
	• Reduce dependence on fossil fuels by using electrical vehicles		X

REFERENCES

Cabeza, L F, Bai, Q, Bertoldi, P, Kihila, J M, Lucena, A F P, Mata, É, Mirasgedis, S, Novikova, A and Saheb, Y, 2022. Buildings, in *IPCC, 2022: Climate Change 2022: Mitigation of Climate Change. Contribution of Working Group III to the Sixth Assessment Report of the Intergovernmental Panel on Climate Change* (eds: P R Shukla, J Skea, R Slade, A Al Khourdajie, R van Diemen, D McCollum, M Pathak, S Some, P Vyas, R Fradera, M Belkacemi, A Hasija, G Lisboa, S Luz and J Malley), (Cambridge University Press: Cambridge). doi: 10.1017/9781009157926.011.

Darby, S and Fawcett, F, 2018. Energy sufficiency – an introduction: A concept paper for ECEEE, Environmental Change Institute, University of Oxford.

Viable and feasible decarbonisation – Tent Mountain coalmine repurposing for the clean energy economy

B D Saffron[1]

1. Executive Advisor, GHD Limited, Calgary, AB, Canada, T2Z0K6. Email: ben.saffron@ghd.com

INTRODUCTION

In 1983, open cut coal mining operations were suspended at the Tent Mountain coalmine, located in Alberta, Canada in an area called Crowsnest Pass.

Montem Resources ('Montem'), an Australian mining company, purchased the asset in 2016. They conducted several drilling campaigns and undertook environmental monitoring with the intent to restart operations. In 2020, Montem completed a Definitive Feasibility Study, which indicated a mine life of more than 14 years.

The location of the Tent Mountain mine and the layout of existing historical operations is provided in Figure 1.

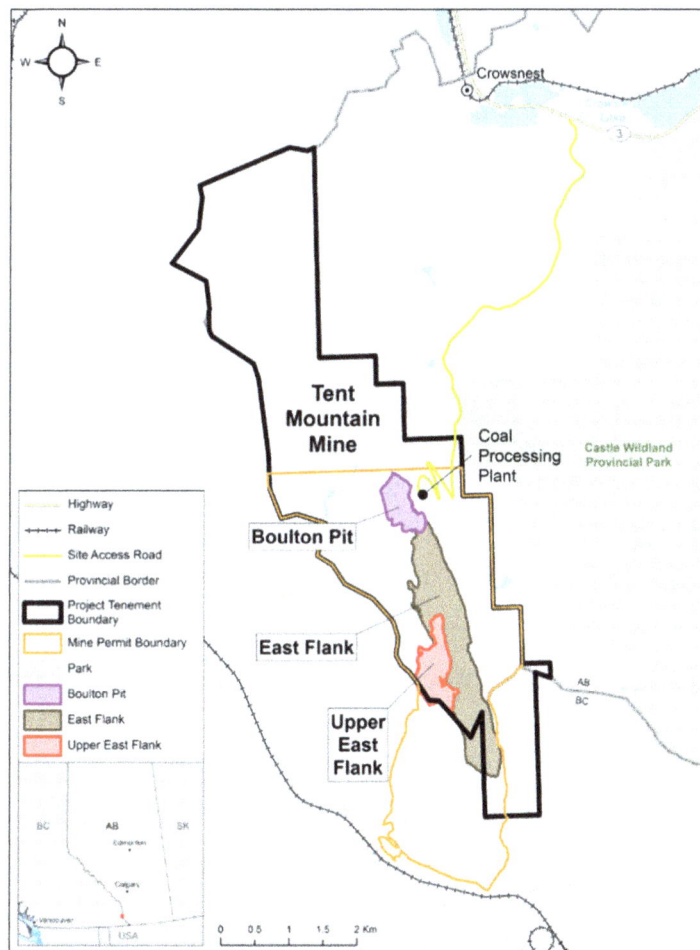

FIG 1 – Location and layout of the Tent Mountain mine site, near Crowsnest Pass, Alberta, Canada (courtesy of Montem).

CHALLENGE

Despite Montem moving the mining restart forward within the Provincial framework at the time, the proposed mining project was ordered to undergo a Federal impact assessment before it could operate (Government of Canada, 2021).

The Federal review process can take many years, with some projects taking over five years and with no guarantee of success. As a result, Montem sought to study alternatives for the Tent Mountain asset that they owned.

The challenge for Montem was to find a way to deliver a sustainable and socially-accepted project with strong economic value that could attract investment and utilise the unique features of the Tent Mountain site and location. The transformation project would need to be both financially viable and technically feasible.

METHOD

To commence an opportunity identification process, Montem engaged multiple professional services firms to look at their unique issue and work collaboratively to identify a technically and financially viable solution.

There are three main ways in which Montem looked to identify the opportunity and move the opportunity forward with the greatest chance of success. These were:

1. Utilise a diverse set of skill sets and knowledge to look at the issue in different ways.

2. Undertake a logical project assessment process.

3. Establish a steering committee of key advisors to move the project forwards.

GHD was consulted initially to look at enhancing an early PHES (Pumped Hydro Energy Storage) concept and preliminary financial returns if it were to be deployed in isolation. This exercise included looking broadly at renewable energy and alternative or clean fuel (including hydrogen) opportunities that might be possible, and driving the opportunity identification process in a collaborative fashion. GHD then took the main role of developing the clean hydrogen portion of the project.

Other firms were tasked with further detailing the PHES scope and return on investment, risks and opportunities, feasibility, and defining the PHES project development, whilst others were engaged on matters of regulatory and permitting requirements, grid interconnectedness, and the important Indigenous aspects of a clean energy project in this location.

Through the bringing together of different groups and initial divergent or 'free' thinking, the conceptualisation is an example of what is possible when multiple, diverse skill sets, and thinking is brought together to solve a compelling challenge.

The steps that resulted in the transformation opportunity were:

1. Framing the issue and brainstorming 'free thinking' options.

2. Value chain interrogation and opportunity definition, including alternatives.

3. Initial market sounding, commercial development and exploratory discussions with prioritised end-users.

4. Initial high-level financial viability assessment through financial analysis and benchmarking.

5. Understanding risks and constraints, and interconnectedness of the project elements.

6. Additional technical and business case studies. Including water balance and testing, geotechnical investigation, siting for the hydrogen facility and water constraint understanding.

7. Understanding the key levers for value, including Indigenous participation, clean fuels standard credits, scaling time frames, key consortia participants, captive end-market users and propensity to pay.

8. Development roadmap with approaches to de-risk the project at various points.

The transformation scheme was deemed by Montem to be both viable and feasible, Montem formed a steering committee led by the previous Chair of the Alberta Electric System Operator ('AESO'), which included subject matter experts from across each of the project elements. This has ensured that the project was aligned across elements and moved forward in an efficient manner.

RESULT

The resulting conversion project is named the Tent Mountain Renewable Energy Complex ('TM-REX') and includes a 320 MW/4800 MWh pumped hydro energy storage facility (PHES), a 100 MW green hydrogen production electrolyser, and a 100 MW wind farm off-site (originally considered on-site) (Montem Resources, 2021; Fletcher, 2021).

The PHES utilises an existing legacy open mining pit called the 'Upper Reservoir' shown in Figure 2, which shows the general layout and elements of the PHES. This Upper Reservoir is presently filled with water and collects rain and snowmelt throughout the year. A Lower Reservoir, approximately 300 m below, would be constructed with a dam structure and existing natural terrain. A powerhouse and surface penstock would complete the required infrastructure, pump water from the Lower to Upper Reservoir when power prices were low and sell power when prices were high.

FIG 2 – Overlay rendering of the PHES configuration at Tent Mountain (courtesy of Montem).

Wind power would be the renewable energy supply for both the PHES and the hydrogen electrolyser. The off-site wind farm is proposed to be developed by the Piikani Nations, an Indigenous group in the region.

'Green' hydrogen is a very low CO_2 intensity fuel when it is created using renewable energy. The other feedstock required is potable water, for which there are several sustainable options in the area. Electrolysers work by splitting water into hydrogen gas and oxygen gas. The hydrogen is then stored at pressure and transported to an end user. For this project, there were a few opportunities identified; the most promising of which are mining haul trucks at other operations and hydrogen locomotives (both which would directly substitute diesel as a fuel) and natural gas blending opportunities.

The full-scale, 100 MW hydrogen production facility will produce around 14 000 tonnes per annum of hydrogen. This would be enough to displace approximately 50 million litres of diesel from large highway trucks, or the equivalent of approximately 200 000 t of CO_2 emissions each year.

The TM-REX development displays several key attributes as described in Table 1.

TABLE 1

Attributes of the Tent Mountain Renewable Energy Complex.

Attribute	Description
Utilises leading edge renewable technology	Using wind farms and water based electrolysers, as a business change enabler to produce zero-emission fuel and power. The 100 MW electrolyser if in operation today would be the largest green hydrogen facility globally
Has a very high societal and environmental impact	In new, green jobs creation in the region, and through reducing transport emissions significantly and pivoting from coal mining to renewable fuels
Involves Indigenous communities completely	Through partnerships for wind farm development and broader participation and joint ownership opportunities
Creates a new business for Montem, and a viable use for this idle asset	To pivot from coal mining to green power, energy storage and fuels as part of their corporate portfolio. Further, there is a new and differentiated opportunity for investors to partake in a unique opportunity
Clean fuel supply of hydrogen as a substitute for diesel	Providing green hydrogen production in this region is the first of its kind and has the potential to aid decarbonisation of multiple industries such as heavy haul transport and other regional coal mining operations
Providing grid stabilising energy storage	Pumped hydro energy storage firms the grid and allows additional renewable energy projects to be added
A first of its kind in North America	Incorporating renewables with pumped hydro and green hydrogen production has not been undertaken in North America
Commercial potential is broad and large	Both for hydrogen sales and off-takers and for the pumped-hydropower system to the grid
Will be an iconic green energy legacy project	Supporting the energy transition well into the future of low carbon energy

As of March 2023, feasibility studies and in-depth business cases have been undertaken, and the project has recently been agreed to undertake joint-venture development with a major utility in Alberta, TransAlta (TransAlta, 2023).

SUMMARY

The TM-REX is a novel approach to re-imagining and then physically re-purposing an idle coalmine. It takes advantage of natural topography and previously disturbed areas, market, and political support for both green hydrogen and a grid stabilising energy storage system.

The TM-REX development is one that achieves many objectives across Indigenous, social, political, environmental, and energy transition elements. Importantly, it is both financially viable and technically feasible decarbonisation.

It is a compelling transformation that is an example of what is possible when multiple, diverse skills and thinking are brought together to solve a complex challenge of transforming a legacy mine.

ACKNOWLEDGEMENTS

The author would naturally wish to acknowledge and thank Montem Resources for their strong and collegiate working relationship, and in particular Montem's CEO and Managing Director, Peter Doyle.

REFERENCES

Fletcher, R, 2021. Instead of a coal mine, this Alberta mountain may now become a 'green energy complex', CBC News, October 20. Available from: <https://www.cbc.ca/news/canada/calgary/montem-tent-mountain-green-energy-complex-1.6218293>

Government of Canada, 2021. Tent Mountain Mine Redevelopment Project, Impact Assessment Agency of Canada. Available from: <https://iaac-aeic.gc.ca/050/evaluations/proj/81436>

Montem Resources, 2021. Tent Mountain Renewable Energy Complex. Available from: <https://montem-resources.com/projects/tent-mountain-renewable-energy-complex/>

TransAlta, 2023. TransAlta Announces Acquisition of 50% Interest in Early-Stage Pumped Hydro Energy Storage Development Project, TransAlta Newsroom. Available from: <https://transalta.com/newsroom/transalta-announces-acquisition-of-50-interest-in-early-stage-pumped-hydro-energy-storage-development-project/>

Getting your cut-off grade policy right for better ESG outcomes

J D Tolley[1]

1. AMC Consultants Pty Ltd, Melbourne Vic 3000. Email: jtolley@amcconsultants.com

ABSTRACT

The shift towards clean energy technologies is having a significant impact on the mining industry, with the demand for minerals such as copper, nickel, cobalt, lithium, manganese, and graphite increasing exponentially. To meet the goals of the Paris Agreement, mineral requirements for clean energy technologies would need to quadruple by 2040, and to achieve net-zero globally by 2050, mineral inputs in 2040 would need to be six times more than today.

The most significant driver of mineral demand is electric vehicles (EVs) and battery storage, with lithium seeing the fastest growth, and copper demand doubling over the same period due to the expansion of electricity networks globally. The demand for these minerals is expected to rise significantly as investments in clean energy technologies continue to grow in the future. As the world moves towards a decarbonised economy, the minerals industry is under increasing pressure to reduce its environmental and social impacts and meet the increased demand for minerals.

Mining companies are facing additional challenges due to the depletion of near-surface mineral deposits globally, with more mines transitioning from open pits to underground. The transition to underground mining requires the right strategy to economically access deeper mineable inventories and to keep up with future clean energy technology demands to achieve net-zero globally by 2050. One key way to achieve this is through decarbonisation using 'green' mining technologies such as replacing current primary diesel mining fleets with battery electric vehicles (BEVs). However, these measures through electrification overlook the bigger opportunity to maximise the socio-economic returns from the mine while also reducing the overall carbon footprint.

Corporate cut-off grade policy is one of the most significant influencers on the total environmental and social impacts of the mining industry. The purpose of a corporate cut-off grade policy is to implement strategies aligned with achieving all the mining companies' corporate goals, with decarbonisation targets becoming a primary driver for mining companies. Table 1 shows the evolution of cut-off theory with the increasing number of dimensions and driving inputs per iteration.

TABLE 1

Evolution of cut-off theory.

Cut-off grade	Date established	Dimension	Driving inputs
Break-even	Pre-1950s	One-dimensional	Financial only (price and cost)
Mortimer	1950	Two-dimensional	Financial and geology (tonnage/grade relationships)
Lane	1964, 1988	Three-dimensional	Financial, geology and system capacities for rock, ore and product
Mine strategy optimisation	Mid-1990s	Multi-dimensional	Financial, geology, system capacities, multiple mines/plants, multiple products, sequencing and timing, variations over time etc

Conventional break-even cut-off grades (BECOGs) are based on simplistic and static metal price and production cost assumptions. BECOGs often result in large quantities of marginal material being mined and processed that can be responsible for significant proportions of total emissions, land disturbance, and an increase in carbon footprint over the life-of-mine. BECOGs don't consider the full range of factors that can impact a mining project, such as ore grades/value, mine capabilities and constraints, or ESG implications and limitations.

Cut-off grades may vary substantially over the life of an operation due to variations in sustaining capital and operating costs, grade distribution, mine and mill performance, metal prices, exchange rates, mining method changes, mining and treatment capacities, resource additions and changing corporate goals. Mine strategy optimisation is a robust and comprehensive assessment of multiple input and scenario combinations to uncover all the strategic options available and is the key to achieving decarbonisation targets while also maximising socio-economic returns.

Figure 1 shows the typical mine value life cycle from discovery>projects>development>operations >decommissioning. The stars coloured green (project) and blue (operation) in Figure 1 shows the stage-gates where the cut-off is a key driver to proceed to the next stage of the mine life cycle, sustain the operation at steady-state production, or reinvest in the operation because of major changes or additional capability requirements. Through each of these stage-gates across the mine value life cycle the impacts on ESGs and the inclusion of decarbonisation initiatives form a crucial piece of the puzzle to unlock the optimal strategy for achieving mining companies' corporate goals and satisfy all key stakeholders involved.

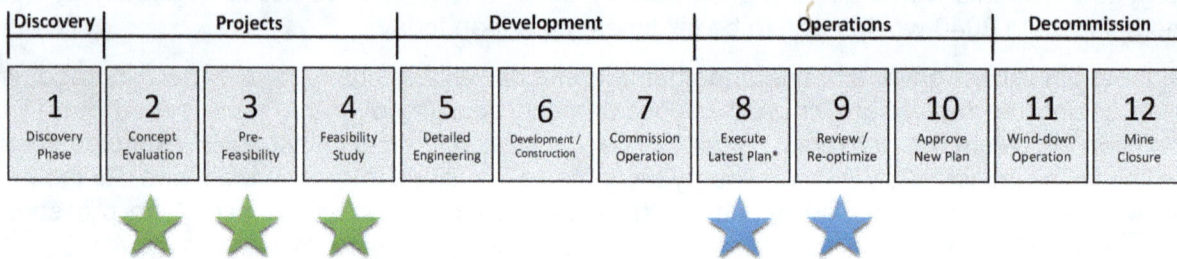

Discovery	Projects			Development			Operations			Decommission	
1	2	3	4	5	6	7	8	9	10	11	12
Discovery Phase	Concept Evaluation	Pre-Feasibility	Feasibility Study	Detailed Engineering	Development / Construction	Commission Operation	Execute Latest Plan*	Review / Re-optimize	Approve New Plan	Wind-down Operation	Mine Closure

Includes ramp-up, steady-state production and ramp-down depending on latest plan.

FIG 1 – Mine value life cycle.

The mining industry and its stakeholders have a great opportunity to get serious about sustainably reducing emissions by addressing this core issue and incorporating the decarbonisation initiatives into the cut-off grade policy and overall mining strategy. This paper describes a more advanced approach to developing the best overall strategy for mining projects, the common flaws in conventional strategic planning and the magnitude of the negative environmental and social impacts this has, and the opportunities for our industry to do better for all stakeholders in the future.

ACKNOWLEDGEMENTS

The author would like to thank A J Hall from AMC Consultants in Melbourne for their assistance in preparing this work.

Developing regional approaches – shifting from operating independently to collaboratively

Maximising regional opportunities for in-pit tailings disposal for Queensland coalmines

K Baker[1] and J Purtill[2]

1. Manager, Office of the Queensland Mine Rehabilitation Commissioner, Brisbane Qld 4000. Email: kate.baker@qmrc.qld.gov.au
2. The Queensland Mine Rehabilitation Commissioner, Office of the Queensland Mine Rehabilitation Commissioner, Brisbane Qld 4000. Email: james.purtill@qmrc.qld.gov.au

INTRODUCTION

One of the key challenges for the mining industry globally is the rehabilitation of mine waste, and Queensland is no exception.

Out of pit waste structures including waste rock dumps and tailings storage facilities, create novel landforms which require ongoing maintenance and pose risks to the environment and sometimes to communities. Where multiple mines co-exist, such as parts of Queensland's Bowen Basin, individual mines can often be constrained by adjoining operations. This presents opportunities for broader rehabilitation strategies to be considered, more than the current single mine focus, that may provide better environmental outcomes for a region.

Given recent global disasters involving tailings storage facilities, and our team's endeavours to drive Queensland to achieve leading practice in mined land rehabilitation, we are exploring the potential to maximise in-pit tailings disposal opportunities for coalmines through shared waste management strategies.

A shared waste management strategy is where two (or more) mines, owned by different companies and operated under different environmental permits, transport tailings from one mine to the other in close proximity, for disposal in a disused pit. This would reduce the number of waste structures to be built, operated, rehabilitated, and managed in perpetuity.

Our objectives are to identify and remove barriers to tailings transfers between sites to:

- minimise environmental risk
- minimise long-term liability of managing mining waste and associated structures
- support improved outcomes for the local community.

To determine the scale of coal waste disposal practices in Queensland, internal data and publicly available imagery was analysed by the Department of Environment and Science in 2022. Approximately 6500 ha of land in Queensland is utilised for coalmine tailings storage facilities or co-disposal areas, with ≈ 3700 ha of this being out of pit.

BARRIERS TO SHARED COALMINE WASTE MANAGEMENT

As there are currently no examples of shared coalmine waste disposal strategies in Queensland, it is necessary to investigate and synthesise if there are barriers to such a strategy.

We grouped potential barriers into three main categories – regulatory, operational and financial. Each of these groups was investigated to understand the nature of their extent and significance.

Regulatory barriers

We investigated the legislative and policy settings that intersect with inter-mine waste transfers. Firstly, we investigated the pathway for shared mine waste disposal opportunities under current Queensland legislation. Following this investigation, further research was undertaken into which pieces of legislation, policies, standards and systems need to be considered and requirements addressed an if inter-mine waste disposal strategy were to be pursued.

Although disposal of waste substances resulting from the winning or extraction of a mineral is included in the definition of a 'mine' in Queensland (Section 6A of the *Mineral Resources Act 1989*), there are overlaps with the *Waste Reduction and Recycling Act 2011* (WRR Act), *Environmental*

Protection Act 1994 (EP Act) and Environmental Protection Regulation 2019 which impact shared mine waste approaches.

The WRR Act sets out the waste management hierarchy for the State, with disposal being the least preferred option. However, where disposal is the only option, the WRR Act doesn't account for the merits of different disposal options. The WRR Act also defines what a waste disposal facility is, waste levy requirements, exemptions, and end of waste codes.

The EP Act sets the environmental regulatory framework for the State, including how environmental authority (EA) permits and progressive rehabilitation and closure plans (PRC plans) are granted and managed. Although no coal mining EAs currently allow shared waste management strategies, the EP Act does not explicitly prevent the approach, and therefore such strategies could be considered.

Environmental Protection Regulation 2019 defines regulated and non-regulated waste and sets out the environmentally relevant activities (ERA's).

Operational barriers

Operational considerations include progressing any required amendments to both mines' EAs, estimated rehabilitation cost payments and possibly PRC plans, if either mine has one approved. Physical operational barriers are harder to define, as they are specific to the sites involved in the shared waste management strategy. However, considerations include: volumes of waste to be accepted, type of waste to be accepted, geotechnical stability, location of infrastructure between sites and any characterisation required to ensure the receiving mine is willing to accept the tailings and inform how it would need to be managed to ensure the protection of environmental values. Considerations also include how the waste would be moved: whether it would need to leave a mining tenure, how it would be physically transported, any impacts to the Site Senior Executive role and requirements, and any internal operational and safety documents that would need updating to give effect to the strategy.

Financial barriers

An investigation of the financial barriers provided the majority are matters for the companies.

One financial barrier in the public domain however is the impact to the mines estimated rehabilitation cost calculations. The source mine for the waste would not have to build and/or operate, manage and rehabilitate an out of pit tailings storage facility, which would have significant impacts to the estimated rehabilitation cost calculations. The receiving mine would also have greater backfill material from the waste and likely have to source less material to backfill the void being used, which may reduce costs and produce a landform that is closer to pre-mining contours than may otherwise be achieved.

Another barrier is in relation to liability. The implications of where the liability for the waste rests has costs, however the liability in itself could also be a barrier for the company who would have to accept that responsibility.

CONCLUSION

Despite the many considerations and current barriers to their adoption, the development of shared waste management strategies between mining operations presents opportunities for significant benefits to the environment and stakeholders through reduced waste structures to be rehabilitated and left in perpetuity to be managed in the region. To help realise the benefits of shared coalmine waste strategies, the Office of the Mine Rehabilitation Commissioner is undertaking further work, with the aim to present a conceptual model of an inter-mine tailings transfer scheme.

ACKNOWLEDGEMENTS

We would like to thank the OQMRC team for their support of the project and their reviews of drafts, notably Louisa Nicolson, Megan Clay and Jason Dunlop.

REFERENCES

Department of Environment and Science, 1994. *Environmental Protection Act 1994* (EP Act), December 1994.

Department of Environment and Science, 2011. *Waste Reduction and Recycling Act 2011* (WRR Act), August 2011.

Department of Environment and Science, 2019. *Environmental Protection Regulation 2019*. August 2019.

Department of Resources, 1989. *Mineral Resources Act 1989*, October 1989.

Developing a regional data catalogue to support the technical assessment of post-mining land use

C M Côte[1], P Bolz[2], L Pagliero[3], M Shaygan[4], M Edraki[5] and P Erskine[6]

1. Director, Centre for Water in the Minerals Industry, Sustainable Minerals Institute, University of Queensland, St Lucia Qld 4072. Email: c.cote@uq.edu.au
2. Research Fellow, Spatial Science Specialist, Centre for Water in the Minerals Industry, Sustainable Minerals Institute, University of Queensland, St Lucia Qld 4072. Email: p.asmussen@uq.edu.au
3. Research Fellow, Hydrologist, Centre for Water in the Minerals Industry, Sustainable Minerals Institute, University of Queensland, St Lucia Qld 4072. Email: l.pagliero@uq.edu.au
4. Research Fellow, Soil Scientist, Centre for Water in the Minerals Industry, Sustainable Minerals Institute, University of Queensland, St Lucia Qld 4072. Email: m.shaygan@uq.edu.au
5. Associate Professor, Group Leader Environmental Geochemistry, Centre for Water in the Minerals Industry, Sustainable Minerals Institute, University of Queensland, St Lucia Qld 4072. Email: m.edraki@uq.edu.au
6. Director, Centre for Mined Land Rehabilitation, Sustainable Minerals Institute, University of Queensland, St Lucia Qld 4072. Email: p.erskine@uq.edu.au

INTRODUCTION

Mines in Queensland must now prepare a Progressive Rehabilitation and Closure (PRC) plan, with the requirement to justify the selection of the post-mining land use (PMLU) for each rehabilitation domain. Given that there are nearly 200 mines who must undertake a comprehensive PMLU assessment, there is a need to develop methods that will minimise the duplication of effort.

As part of a broader initiative on innovative post-mining land use planning, this project undertook a regional analysis of the technical suitability of several PMLUs.

METHODS

The methodology involved:

- Identifying potential PMLUs based on an analysis of the Australian Land Use and Management (ALUM) classification (ABARES, 2016), a review of international and national practices and of technical advances in agriculture, horticulture, aquaculture, residue management, water treatment and renewable energy production that could be incorporated to support or create innovative PMLUs.

- Gathering baseline information about the spatial distribution of biophysical parameters and infrastructure networks in Queensland.

- Performing spatial analyses to derive an assessment of the technical suitability of each identified PMLU.

Spatial data

Assessing the technical feasibility of a PMLU required compiling information about:

- Climate, to assess the potential to sustain vegetation establishment and resilience, cropping systems and production of renewable energy.

- Soil types, to determine the land capability and the suitability for agricultural systems, such as cropping and grazing.

- Water resource planning and supply mechanisms, which are essential for some PMLUs, most notably irrigated cropping systems.

- Infrastructure: transport corridors (roads and rail), electricity supply network, gas and water pipelines. For instance, a renewable energy project will require access to transport infrastructure for construction and maintenance and to the electricity grid to export its production.

- Geology, as it dictates the type or category of waste materials generated by mining activities, which will impact on the quality of the water that comes in contact with it. It can also provide an indication of the potential for waste to contain valuable elements thereby providing an economic activity from waste reprocessing.

The Queensland Spatial Catalogue (QSpatial, https://qldspatial.information.qld.gov.au) is the central repository for spatial data in Queensland. Complemented by climate data from the Bureau of Meteorology, it provides extensive data sets that characterise the biophysical characteristics governing PMLU suitability. These were compiled to provide a regional data catalogue that can support PMLU assessment (Figure 1).

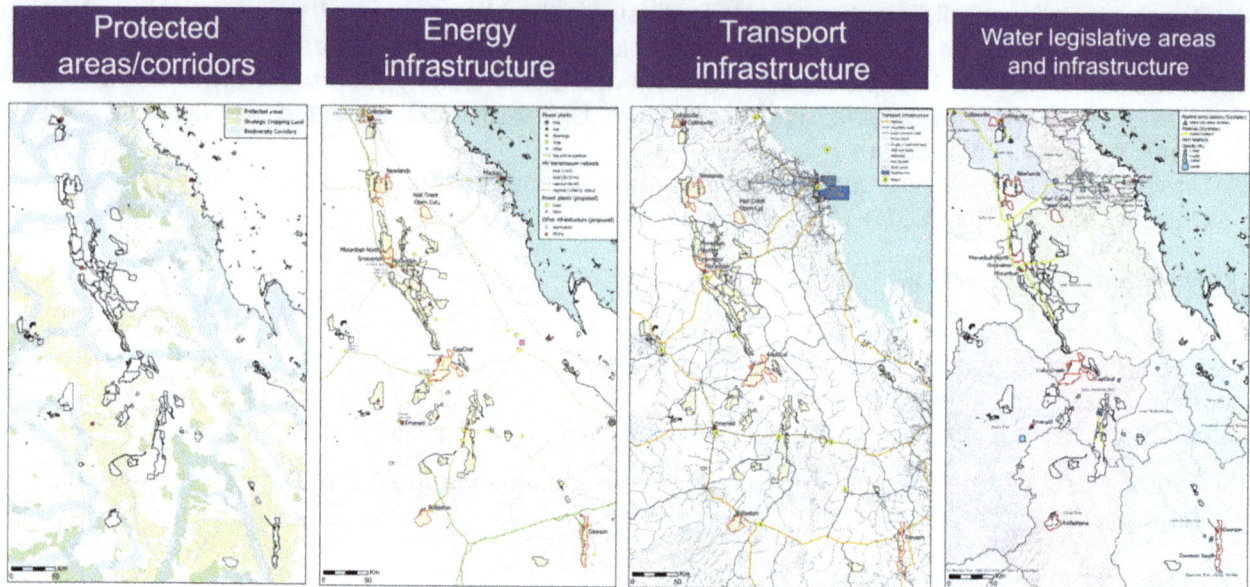

FIG 1 – Extract from the regional data catalogue created for the Bowen Basin.

Assessment of technical suitability

Potential PMLUs that were identified included: native ecosystems, grazing, cropping, renewable energy (solar and wind), phytomining (the harvest of metals from plants known as hyperaccumulators), protected horticulture (the re-purposing of mine buildings to grow food products), intensive livestock, manufacturing and tourism.

GIS-based multicriteria decision-making analysis is a common approach to assess the suitability of a land use (eg Malczewski, 2004). A simplified approach was adopted here, where for each PMLU, suitability indicators were derived (Table 1). For instance, grazing would only be suitable if the soil could support pasture, precipitation was consistent, and meat could be exported via a transport network.

Spatial analyses were undertaken to quantify these indicators. Qualitative rankings were then derived based on the value of the indicators.

TABLE 1

Indicators selected to assess the technical feasibility of PMLUs.

Potential PMLU	Indicator 1	Indicator 2	Indicator 3
Native ecosystems	Ratio of mine lease area to national park and protected areas	Ratio of mine lease area to area covered by biodiversity corridors	Proximity to remnant ecosystems that are not part of the formal protected areas network but near the mine lease
Grazing	Soil suitability	Precipitation	Proximity to transport Infrastructure
Cropping	Soil suitability	Precipitation	Proximity to transport Infrastructure
Solar Energy	Solar radiation	Proximity to transmission network	Proximity to potential collaborators
Wind Energy	Wind speeds	Proximity to transmission network	Proximity to potential collaborators
Phytomining	Likelihood of trace elements in waste	Likelihood of trace elements in mine water	Presence of hyper-accumulators in region
Protected horticulture Intensive livestock	Likelihood of access to water	Proximity to transport Infrastructure	
Manufacturing	Proximity to transport Infrastructure		
Tourism	Proximity to transport Infrastructure (Airports)		

RESULTS

An example of the ranking process is presented in Table 2, where 'very suitable' is shown in dark green, 'suitable' in light green, 'possible' (but often with additional management strategies) in yellow and 'present challenges' in orange.

TABLE 2

Example of qualitative ranking of PMLU suitability.

Potential PMLU	Mount Isa	Weipa	Charters Towers	Bowen Basin	Surat Basin
Native ecosystems					
Grazing					
Cropping					
Solar Energy					
Wind Energy					
Phytomining					
Protected horticulture					
Intensive livestock					
Manufacturing					
Tourism					

Dark Green: very suitable; Light Green: suitable; Yellow: possible (but often with additional management strategies); Orange: present challenges.

Rankings can be used to select the most technically feasible PMLUs for each area. The intent is that these short lists can facilitate discussions and minimise effort when assessing the potential for a PMLU to be included in a PRC plan. As the catalogue is based on regional data sets, additional analysis will be required at a site scale to reflect local conditions.

Moreover, rankings can be used to provide an overview of the strengths and constraints of each region with respect to its potential for supporting PMLUs. With current practices, selection of PMLU is undertaken on a site-by-site basis. The regional context is not considered, and this could lead to missed opportunities.

To our knowledge, there has been no other attempt at including regional biophysical context in the assessment of PMLU. This regional PMLU assessment methodology can be applied to any mining region, if a repository of spatial data is available.

ACKNOWLEDGEMENTS

The project was funded by the Queensland Resources Council (QRC) and the International Council on Mining and Metals (ICMM). It was overseen by a working group of industry representatives. We wish to acknowledge the contributions of the working group members who provided direction to the project, as well as continuous feedback.

REFERENCES

ABARES, 2016. The Australian Land Use and Management Classification – Version 8. Australian Bureau of Agricultural and Resource Economics and Sciences, Canberra, CC by 3.0. Available from: <agriculture.gov.au/abares/publications>

Malczewski, J, 2004. GIS-based land-use suitability analysis: a critical overview, *Prog Plan*, 62:3–65. https://doi.org/10.1016/j.progress.2003.09.002

Initiating collaborative planning for post-mining land use in Victoria's Latrobe Valley

T Foran[1] and J Reeves[2]

1. Senior Research Scientist, CSIRO, Canberra ACT 2614. Email: tira.foran@csiro.au
2. Associate Professor, Institute of Innovation, Science and Sustainability, Federation University, Gippsland Vic 3841. Email: j.reeves@federation.edu.au

INTRODUCTION

We report on factors motivating the emergence of a 2022–2024 initiative to enhance multi-stakeholder planning around visions and options for post-mining land use (PMLU) in Victoria's Latrobe Valley, where three brown coalmines are closed or slated for closure within 12 years (ENGIE Hazelwood mine and power station, 2017; Energy Australia Yallourn mine, 2028; and AGL Loy Yang mine, 2035). Significant focus has been devoted to economic transition support for mine and power station workers (Weller, 2019). By contrast, processes by which options for post-coal era economic development of 130 km² mine land are perceived by stakeholders as insufficient (Foran, Barber and Ackermann, 2022) (the 'post-coal era' refers to economic activities at a period of time after the decline of coal-fired electricity generation in Victoria. Leaves open the possibility of coal mining for hydrogen production but regards it as a niche activity). A shared PMLU development vision does not exist to guide and mobilise rehabilitation planning and economic development investment for the Valley. This paper introduces the conceptual designs of the 2022–2024 'Latrobe collaborative planning' ('LCP') initiative. We then propose potential 'drivers' of collaboration among the LCP's participating stakeholders.

CONCEPTUAL DESIGN

LCP is based on two conceptual frameworks related to governance and policy implementation. First, drawing on a *policy regime* framework (May and Jochim, 2013), we argue that the Valley's post-coal mining era transition challenges can be conceptualised as a challenge of developing a policy regime which can span two existing domains: mine closure planning; and regional development (Foran, Barber and Ackermann, 2022). Second, drawing on an integrative framework for *collaborative governance* (Emerson, Nabatchi and Balogh, 2012) we argue that diverse stakeholders can work together to craft components of an improved policy regime. As shown in Figure 1 a policy regime consists of *policy arguments* on how to realise various values in society; *institutional arrangements* which channel attention more or less effectively to act in accordance with those arguments; and systemic effects (notably stakeholder support and opposition) as implementation unfolds over time.

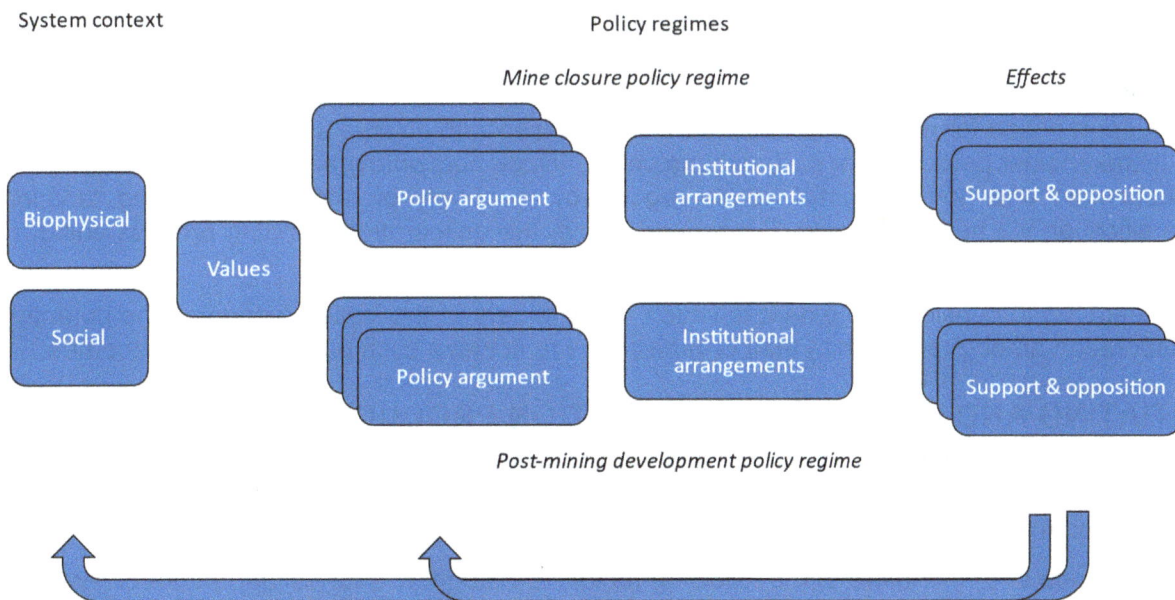

FIG 1 – Two policy regimes relevant to post-mining transition.

Foran, Barber and Ackermann (2022) argued that mine closure planning and post-mining regional development in Australia are two overlapping but distinct policy regimes. They argued that the societal value of *risk mitigation* appears to be regarded as of supreme importance from the perspective of the mine closure regime. However, from the perspective of the post-mining development regime, they are regarded as of high importance, alongside values of *pragmatism* and *adaptiveness* (Foran, Barber and Ackermann, 2022: section 7.2).

The mine closure planning policy regime in Victoria distributes authority and responsibility among multiple government agencies and private parties. The prevailing arrangements under the Mineral Resources (Sustainable Development) ['MRSD'] Act 1990 consist of statutory planning governed by a proponent-led model, in which individual licensees propose post-mining landforms and closure criteria as part of 'declared mine rehabilitation plans'. Proponents are expected to conduct community consultation around implications for PMLU of their post-mining landforms.

Mine closure plans must also conform to requirements of Victoria's Planning and Environment Act, its Environmental Effects (EE) Act, and potentially the federal EPBC Act. ENGIE, the first operator to close, is generating an Environmental Effects Statement under the Environmental Effects (EE) Act, as well as under the EPBC Act.

Proponents are also expected to follow guidelines for regional coordination in the 2020 Latrobe Valley Regional Rehabilitation Strategy (LVRRS). The LVRRS is described by legal scholars as an 'integrative' policy instrument under the MRSD Act (Gardner *et al*, 2022). It is an evolving policy and planning initiative, coordinated by agencies responsible for earth resources regulation and for water (within Department of Energy, Environment and Climate Action since January 2023) (an independent agency, the Mine Land Rehabilitation Authority (MLRA), was established in 2020 to perform monitoring and evaluation functions, providing assurance to the public that licensees and public agencies are planning for rehabilitation. A related duty of MLRA is to promote participation of Latrobe Valley community and stakeholders in implementation of the LVRRS). One intent of the LVRRS is to provide an overarching framework for ensuring that values of the wider community are part of rehabilitation planning. This intent is expressed in a series of principles and guidelines related to use of water resources, acknowledging other users and uses as well as impacts of a drying climate. Likewise, the Strategy explicitly supports identifying practicable and intended land use for rehabilitated landforms (DJPR and DELWP, 2020; Gardner *et al*, 2022).

However, the LVRRS has not yet designed a mechanism (ie policy instrument) by which community values related to PMLU could be integrated with concerns about biophysical safety and stability (values at the core of mine closure planning), yielding options for consideration and eventual choice by Traditional Owners, the community, licensees, and government. The lack of such mechanism is a weakness in *institutional arrangements*.

The LCP initiative responds to the above gap in institutional arrangements. It aims to design and test a mechanism involving a multi-stakeholder deliberation on preferred post-coal-era development options for mined lands of the Latrobe Valley. The deliberation would consider financial, economic, environmental, social and cultural impacts of alternative options. Participatory multicriteria analysis will support option formulation. If proposed and assessed with diverse and rigorous community and multi-stakeholder participation, such options could include noteworthy and potentially novel bundles of economic and non-economic values (eg improving supply of affordable housing; economic diversification via new industries). If it proves viable, the deliberative planning process could form part of the policy mechanism currently missing.

This leads us to consider the extent to which stakeholders of the Valley recognise the potential of collaboration among (and within) their organisations to achieve such path-breaking outcomes.

INITIATING AND MAKING COLLABORATION PRODUCTIVE

The LCP is supported by more than a dozen government and private sector partners, including three mine licensees, a power station operator, regulators, local government, Traditional Owners, and multiple community organisations.

The presence of one or more of the following factors is thought to spur collaboration (Emerson and Nabatchi, 2015a):

- uncertainty
- interdependence
- consequential incentives
- initiating leadership.

These drivers may emerge from a 'system context' – a set of social and biophysical elements whose interaction gives rise to a problem/s of public concern (Emerson and Nabatchi 2015a; Foran and Yuen, 2021). Table 1 is a distillation of our experience communicating with diverse stakeholders to develop the LCP initiative, and illustrates how drivers of collaboration may arise from Latrobe Valley's system context.

TABLE 1

Potential drivers of collaboration for mine rehabilitation in Latrobe Valley.

Elements of system context	Driver of collaboration
Values: licensees, government, and community desire net-positive post-coal-era rehabilitation and transition outcome	*Consequential incentives* among *interdependent* licensees, governments, and community to generate and elaborate a shared PMLU vision, surpassing business-as-usual outcomes
Institutional arrangements: core values of mine closure regime (landform safety and stability) -> demand for water for rehabilitation in context of *fully allocated river system water* with *competing actors/values*	Stakeholder contestation and policy *uncertainty* over licensee ability to access river system water for rehabilitation -> *consequential incentives* for licensees and other stakeholders to justify net-positive benefits of PMLU associated with water-based rehabilitation.

A final driver of collaboration – initiating leadership – is arguably provided by the LCP initiative and its cash and in-kind resource partners, who share an interest in values-based analysis and deliberation.

As noted above, the LCP initiative will organise a deliberation on preferred development options for (post-coal-era) mined land, considering impacts of alternatives. The deliberation seeks to identify options preferred by multi-stakeholder participants, and will invite key responsible parties (government agencies and proponents) to respond to those recommendations.

Delivering such a process will require collaboration to be productive. Two components of productive collaborative dynamics are 'principled engagement' and 'shared motivation' (Emerson and Nabatchi, 2015a). In essence the former refers to qualities of reasoned communication, and the latter to qualities of inclusive, legitimating relationship among participants.

The two components have synergistic potential. For example, if mine licensees, government agencies, civil society, and other private sector organisations accept others' legitimate interests, they should be able to agree on an initial set of post-closure landforms and land classifications from which alternative PMLU scenarios can be defined and evaluated ('land classifications' refers here to classification with restrictions on post-closure development as a function of proximity to pit voids and other features deemed risky).

As of June 2023, the initiative is working to facilitate agreement among licensees and other core stakeholders on units, scope, and methods of analysis. The degree to which such agreement can be reached serves as an indicator of collaborative dynamics (Emerson and Nabatchi, 2015b).

In closing, the challenges of post-coal-era transition are deep seated. For us, a meaningful win consists of a co-produced regional planning framework, enabling a stakeholder shared vision, in which feasible mine rehabilitation realises diverse values related to sustainable post-mining development.

REFERENCES

Department of Jobs, Precincts and Regions (DJPR), and Department of Environment, Land, Water and Planning (DELWP), 2020. Latrobe Valley Regional Rehabilitation Strategy, vol 1, State of Victoria, DJPR and DELWP, Melbourne.

Emerson, K and Nabatchi, T, 2015a. *Collaborative governance regimes* (Georgetown University Press: Washington, DC).

Emerson, K and Nabatchi, T, 2015b. Evaluating the Productivity of Collaborative Governance Regimes: A Performance Matrix, *Public Performance and Management Review,* 38:717–747.

Emerson, K, Nabatchi, T and Balogh, S, 2012. An Integrative Framework for Collaborative Governance, *Journal of Public Administration Research and Theory,* 22:1–29.

Foran, T and Yuen, E, 2021. Defining success for CRC-TiME: A collaborative governance approach to impact and evaluation (CRC TiME Limited: Perth).

Foran, T, Barber, M and Ackermann, F, 2022. Understanding the values of stakeholders in Australian post-mining economies (CRC TiME Limited: Perth).

Gardner, A, Poletti, E, Downes, L and Hamblin, L, 2022. Rehabilitation of the Latrobe Valley Coal Mines – integrating regulation of mine rehabilitation and planning for land and water use (CRC TiME Limited: Perth).

May, P J and Jochim, A E, 2013. Policy Regime Perspectives: Policies, Politics and Governing, *Policy Studies Journal,* 41:426–452.

Weller, S A, 2019. Just transition? Strategic framing and the challenges facing coal dependent communities, *Environment and Planning C: Politics and Space,* 37:298–316.

From 'closing a mine' to 'transitioning a community' – evolving from specific post-mining land-uses to a holistic post-mining vision

A Maier[1], M Goldner[2] and C Gimber[3]

1. Senior Consultant, ERM, Melbourne Vic 3000. Email: amanda.maier@erm.com
2. Principal Consultant, ERM, Brisbane Qld 4000. Email: martine.goldner@erm.com
3. Partner, ERM, Brisbane Qld 4000. Email: chris.gimber@erm.com

INTRODUCTION

Given the ongoing global crises of climate change and nature loss, with the World Economic Forum (WEF) identifying them as the top risks to global economies in terms of both likelihood and impact in the coming ten years (WEF, 2022), there is increasing, pressing demand for companies to demonstrate they understand their climate and nature related risks and look toward contributing genuine beneficial environmental outcomes. Within the mining industry, mine closure is the obvious field in which climate and nature positive strategy can be linked with operations. To do so, there is a need to challenge long-held assumptions around closure in the mining and regulatory context; particularly, to shift the focus from closing individual mines to considering regional opportunity generation and value creation.

CONTEXT

It is no longer acceptable to consider mine operations and closure within the narrow boundaries of a mining lease and the life-of-mine (LoM). The world is currently experiencing the dual crises of climate change and nature loss, which are inextricably linked. The shift to a low carbon economy requires a shift to a nature positive economy as well, and this shift is set to drive a significant increase in demand for transition materials like copper, nickel, lithium, and rare earth elements. While critical to the transition, the mining industry itself also contributes to climate change and nature loss, through land clearing and water use, greenhouse gas emissions, contamination and enabling the invasion of alien species. Equally, these crises expose the mining industry to several significant material risks (WEF, 2022).

The mining sector is evolving, with the emergent priority to ensure capital investment is better aligned to environmental, social and governance (ESG) outcomes. With its long-term investment profile and presence in institutions, communities, and significant ecosystems, the mining industry is perfectly positioned to demonstrate environmental and social stewardship through proactively addressing their ESG risks, identifying opportunities, and assuming a leading role in the application and uptake of climate and nature-based solutions, ultimately leaving impacted landscapes and communities with a sustainable, nature positive future. To do so will require novel ways of thinking, operating, and implementing solutions that were not considered even five or ten years ago, and accordingly, both mine operators and regulatory bodies are required to evolve in response to the increasing ESG demands.

CURRENT SITUATION IN MINING

Integrated mine closure planning is critical for a mine's social license to operate (Dzakpata *et al* 2021; Garcia, 2008). In general, existing regulatory frameworks for mine closure focus solely on harm minimisation and rehabilitation, with limited scope for consideration of social aspects and alternative uses for the sites (Vivoda, Kemp and Owen, 2019; Kretschmann, 2020; Beer *et al*, 2022). Whilst regulations vary across jurisdictions (both internationally and within countries), a common overarching objective for mine rehabilitation is to achieve safe, stable, non-polluting, and self-sustaining ecosystems capable of supporting an agreed post-mining land use (PMLU) (Kragt and Manero, 2021). Completion criteria are the metrics used to measure progress towards achieving this objective, and often mine sites need to demonstrate these criteria have been met before the site can be relinquished. However, despite increasing recognition that the success of closure programs is dependent on having a robust and holistic post-mining vision for the site, which in turn is based on ongoing community involvement and support, there is limited mine closure success seen in Australia and across the globe (Kragt and Manero, 2021).

Two contributing factors to this:

1. The PMLU is rarely considered beyond the mining lease and adjacent land uses, with the associated completion criteria often unrealistically defined (eg to achieve full ecosystem restoration, which may not be biophysically possible on a mined landscape).

2. Ill-adapted regulations and incentives, geared toward closure and relinquishment and not on value creation and more interconnected PMLU opportunities.

This can stifle innovation and limit the adoption of forward-thinking land uses, as sites are driven towards the simplest and easiest completion criteria to be able to relinquish the site.

MOVING FORWARD

Reframing completion criteria may be an effective way of overcoming these problems. Rather than focusing on the mining lease and setting prescriptive criteria targeting a local PMLU, criteria could instead be developed that establish a post-mining vision for a region. This should think 'beyond the fence', and consider the broader regional, national, and potentially even international, socio-economic, and environmental context. A post-mining vision keeps options open for the community, caters for water and vegetation in the landscape as a resource in and of itself, and allows for different development opportunities as new technologies arise and/or advance in the future that cannot necessarily be identified during mining or even at the time of mine closure.

To shift the focus from closing individual mines to creating opportunities within a regional context, companies can start by (and indeed, some already are):

- Identifying potential PMLU opportunities and value creation across the regions they operate, through the ESG lens (eg renewable energy projects, habitat corridors, carbon farming etc).

- Identifying existing ESG projects in the regions they operate and how the identified PMLU opportunities may interconnect with and/or be leveraged from existing initiatives, to result in added value and cumulative impact towards nature and climate solutions (eg applying system dynamics theory, such as ecosystem modelling, to understand relationships, interactions, and consequences within a system).

- Collaborating with regional / state / national / international institutions and communities to undertake the above assessments and identify funding opportunities at the regional / state / national / international level.

- Developing site specific suitability criteria (eg physical factors such as topography, solar irradiation, flood risk etc; social factors such as traditional owner groups, population densities, proximity to residential homes, land use zonings etc) for the various PMLUs identified in the wider regional assessment, to drive appropriate and adequate data collection for the site which can then be used to support both decision-making around what PMLU may be appropriate, and understanding a site's climate and nature related risks.

CONCLUSION

The repurposing of mines and social transition are concepts that have recently arisen in the mining industry but lack appropriate regulatory guidance (Hamblin, Gardner and Haigh, 2022). There is a need to challenge base (or long-held) assumptions around closure in the mining and regulatory context, particularly to enable novel post-mining visions to be adopted for regions in which mines operate. Given the changing investor landscape on ESG matters (World Bank, 2021), and with governments worldwide supporting climate and nature related disclosure frameworks (with many becoming mandatory), re-imagining closure as opportunity generation and value creation will benefit the communities (who are directly impacted), the companies (as they attract investors and improve their social license to operate) and help achieve net zero ambitions.

REFERENCES

Beer, A, Haslam-McKenzie, F, Weller, S, Davies, A, Cote, C, Ziemski, M, Holmes, K and Keenan, J, 2022. *Post-mining land uses* (CRC TiME Limited: Perth).

Dzakpata, I, Qureshi, M, Kizil, M and Maybee, B, 2021. *Exploring the Issues in Mine Closure Planning* (CRC TiME Limited: Perth).

Garcia, D, 2008. *Overview of International Mine Closure Guidelines*, presented at the Third International Professional Geology Conference (American Institute of Professional Geologists).

Hamblin, L, Gardner, A and Haigh, Y, 2022. *Mapping the Regulatory Framework of Mine Closure* (CRC TiME Limited: Perth).

Kragt, M and Manero, A, 2021. Identifying industry practice, barriers and opportunities for mine rehabilitation completion criteria in western Australia, *Journal of Environmental Management*, vol 287.

Kretschmann, J, 2020. Post-Mining – a Holistic Approach, *Mining, Metallurgy and Exploration*, 37:1401–1409.

Vivoda, V, Kemp, D and Owen, J, 2019. Regulating the social aspects of mine closure in three Australian states, *Journal of Energy and Natural Resources Law*, 37(4):405–424.

World Bank, 2021. The Changing Wealth of Nations 2021: Managing Assets for the Future. Available from: <https://openknowledge.worldbank.org/server/api/core/bitstreams/c0debd7c-e47a-5213-9959-64659494f791/content>

World Economic Forum (WEF), 2022. *The Global Risks Report 2022,* 17th Edition. Available from: <https://www3.weforum.org/docs/WEF_The_Global_Risks_Report_2022.pdf>

The perfect storm – mine closure in the Latrobe Valley Victoria

A Scrase[1] and J Brereton[2]

1. Technical Director, MLRA, Morwell Vic 3821. Email: antonia.scrase@mineland.vic.gov.au
2. CEO, MLRA, Morwell Vic 3821. Email: jennifer.brereton@mineland.vic.gov.au

INTRODUCTION

The Latrobe Valley is located approximately 150 kms east of Melbourne. Within a very small area there are 3 large brown coalmines and associated power stations, 3 large towns, and numerous smaller towns in the vicinity.

The mines have been generating power for Melbourne and surrounds for the last 100 years, with two of the mines being owned and operated by the Victorian Government (State Electricity Commission of Victoria) prior to privatisation in the 1990s. All three mines were privatised with the expectation that the mine voids would be filled with water to become pit lakes, as the final landform/rehabilitation solution (Hazelwood Mine Fire Inquiry, Teague and Catford, 2016). It could be argued that until relatively recently, this was largely expected, by government and industry, to be the rehabilitation solution for managing the two key risks associated with the sites:

- The geotechnical stability of the mine pits.
- The prevention of fire – through covering coal batters with water.

Transitioning away from mining economies requires collaborative, regional, innovative thinking with a multitude of stakeholders, from a wide range of government departments, industry, First Nations people and broader community. Closure planning and rehabilitation activities require a range of skills and experience, such as, closure planning, various scientific and engineering disciplines (pit water quality modelling, geochemical modelling, cover design, landform design, contaminated site etc), civil engineering, demolition etc. Some of these skills are similar, such as financial modelling, stakeholder engagement, contract and project management. But the majority are different to those required during the operational phase of a mine. These closure skills can also be lacking within government, with Departments more familiar with monitoring and compliance activities associated with operational facilities.

Some key aspects that appear to be critical to planned successful closure are:

- Collaborative regional and localised planning, with industry, government, community, First Nations and other industry, business leaders and entrepreneurs all involved in providing a smooth transition away from mining.
- Well-defined and known regulatory pathways to achieve mine licence relinquishment assists in encouraging investment.
- Appropriately skilled and experienced government and industry personnel.
- Mining companies, who are committed to closure planning and provisioning from the earliest point in project planning.
- Mine employees who are appropriately skilled and experienced in mine closure planning and implementation with access to all the relevant technical expertise required for the project.

Areas like the Ruhr Valley in Germany provide good some examples of smooth processes that facilitate mine economies transition into new economies (World Resources Institute (WRI), 2021).

THE ADVENT OF THE PERFECT STORM

The Latrobe Valley is now facing the potential for the 'perfect storm':

- All three coalmines closing within a relatively short time frame (Hazelwood closed in 2017, Yallourn to close in 2028 and Loy Yang in 2035).
- No clear post mining vision for the Latrobe Valley region.
- No clear pathway to achieve a transition from a mining economy to sequential land uses that support the community and environment.

- A lack of mine closure skills and experience in industry and government.
- A regulatory environment rapidly changing, a lack of regulatory understanding, available guidelines, or certainty in the processes for achieving mine closure.

Left unchecked the perfect storm could result sub-optimal outcomes for the Latrobe Valley, which could include the following issues:

- One or more mine pits unable to be used for other land uses, fenced off, requiring in perpetuity, management of surface and groundwater, geotechnical and erosional stability, fire risks and site security.
- One or more licensees walking away from the mine license before achievement of mine rehabilitation, possibly leaving the Victorian community with ongoing financial impacts.
- Opportunity loss of using the land for other purposes, assisting in the Latrobe Valley losing opportunities to diversify its economy and increased employment.
- Large areas of the Valley being visibly in disrepair with crumbling infrastructure, and social decline in the communities due to loss of opportunities associated with the land areas.

KEY DISCUSSION POINTS

Key aspects that could have an impact on being able to navigate the perfect storm and achieve good rehabilitation outcomes for the Victorian community are:

- The Victorian Mining Legislation and other legislative processes that interact: the paper looks at the past and current mining legislation, what the gaps might be to achieving good outcomes. It acknowledges the crossover with other legislation and the increased complexities of this.
- The Regulators (experienced, well-resourced assessment officers and regulators): the paper discusses the change in approach within Victoria to a more outcome, risk-based focus from prescriptive tick box approach, and suggests that support is needed for the change in workload.
- The Latrobe Valley Regional Rehabilitation Strategy (reviewed and updated to provide a roadmap): the paper notes the input of the Latrobe Valley Regional Rehabilitation Strategy and what has been provided to date.
- The Latrobe Valley declared mine licensees: the paper examines the past, present and progression of closure planning within the Latrobe Valley, looks at some of the key problems for the licensees and the potential outstanding issues.
- Community and Traditional Owners engagement and input into the mine closure and rehabilitation planning processes and implementation (not discussed within this paper).

MOVING FORWARD IN THE LATROBE VALLEY

There is significant, collaborative work progressing in the Valley, with the aim of achieving good outcomes for the region and the Victorian Community:

- Government is commencing the development of guideline documents which will assist with the development of the Declared Mine Rehabilitation Plans (DMRPs).
- The Mine Land Rehabilitation Authority (MLRA) is/will be part of the technical advisory group on developing the guidelines for the DMRP.
- The mine licensees are working towards proposing the final landforms in the DMRPs. Engagement is ongoing with Government regarding the availability of water for the final landform rehabilitation.
- The Mine Licensees are working towards developing a site-specific closure process, including objectives, risk assessments and closure criteria and drafting their DMRPs, which have to be submitted by June 2025.
- Mine Licensees are undertaking or preparing for the potential to undertake an Environmental Effects Statement (EES).
- The MLRA is undertaking a regulatory gap analysis and providing questions back to Government to address.

- The MLRA, mine licensees and other public bodies are regularly engaging on key aspects.
- Research into key areas of mine closure is underway by Cooperative Research Committee Transformation in Mining Economies (CRC TiME) supported by the MLRA, the Latrobe Valley Declared Mine Licensees, Federation University and Government.

HOW TO PREVENT THE STORM FROM DEEPENING

The Latrobe Valley's long history of mining and the incidents described above, give us an indication of what will happen, physically, if the mines are left without suitable rehabilitation. the subsequent outcome is likely to be large areas of derelict land in the Latrobe Valley, unavailable for reuse and generation of economic outputs, with likely significant environmental impacts both in the immediate area and downstream. Full pit lakes provide a viable long-term solution to mitigate the key risks of instability and fire for industry and Victorian community. However, any rehabilitation option comes with consequences, to the environment, to the community, and financial to industry. There is no option that has no consequences. The LVRRS was set-up to develop a regional strategy, not mine specific, and consider common issues, such as regional planning, regional impacts, and access to water and restrictions that may be required to manage the risks to water availability for the region.

Addressing key risk areas identified in this paper could assist in a smoother transition away from mining, and may lower costs and risk to industry, government, the Victorian community and the MLRA. Current activities by all parties need to continue at a pace and consideration given to the following:

Government

- Improving the capacity and capability of Government.
- Streamlining and strengthening the Mineral Resources (Sustainable Development) Act 1990, Regulations and Guidelines.
- Streamlining relevant interacting Acts regulatory processes (Mineral Resources (Sustainable Development) Act 1990, EP Act Environment Protection and Biodiversity Conservation Act 1999 etc), One approval pathway and decision-maker.
- Regional planning developed.
- Reviewing compliance monitoring and reporting throughout the mine life.
- Develop open and transparent processes.

Industry

Depending on where in the life cycle of mine, the mine licensees could consider:

- Improve the closure capacity, capability and experience of the workforce.
- Incorporate closure planning into the business planning, noting that with such short time frames it may not be possible.
- Develop post closure fund.
- Resource the wider supporting skills accordingly: legal, finance, demolition, procurement, contract and project management teams.

REFERENCES

Hazelwood Mine Fire Inquiry, Teague, B and Catford, J, 2016. Hazelwood Mine Fire Inquiry Report 2015/2016 Volume IV – Mine Rehabilitation, 247 p. (Victorian Government Printer: Melbourne). http://hazelwoodinquiry.vic.gov.au/wp-content/uploads/2015/09/Hazelwood-Mine-Fire-Inquiry-Report-2015–2016-Volume-IV-%E2%80%93-Mine-Rehabilitation-web.pdf

World Resources Institute (WRI), 2021, April 1. Germany: The Ruhr Region's Pivot from Coal Mining to a Hub of Green Industry and Expertise, World Resources Institute. https://www.wri.org/update/germany-ruhr-regions-pivot-coal-mining-hub-green-industry-and-expertise

Shifting post-mining land use planning from mine site to regional scale

S Worden[1], K Svobodova[2] and C M Côte[3]

1. Research Fellow, University of Queensland, St Lucia Qld 4072. Email: s.worden@uq.edu.au
2. Marie Skłodowska-Curie Research Fellow, Georg-August-Universität, Göttingen, Germany; Honorary Research Fellow, University of Queensland, St Lucia Qld 4072. Email: kamila.svobodova@uni-goettingen.de
3. Director, Centre for Water in the Minerals Industry, Sustainable Minerals Institute, University of Queensland, St Lucia Qld 4072. Email: c.cote@uq.edu.au

INTRODUCTION

Mines in Queensland are transitioning to the progressive rehabilitation and closure plan (PRCP) framework. Mine operators are required to select and justify appropriate and viable post-mining land uses (PMLUs) following consideration of the surrounding landscape, community views and the objectives of local and regional planning strategies. The new approach offers opportunities for considering innovative PMLU options and collaborative use of post-closure mine assets, such as infrastructure (power, road/rail, buildings), dams and residual voids.

Consideration of regional planning strategies will require a shift in thinking as mine closure is regulated and planned on an individual site basis and selection of PMLUs are commonly considered from this perspective. Planning at regional rather than site-specific scale offers strategic advantages, including the ability to align PMLU selection with the regional context and stakeholder aspirations (which could foster sustainable outcomes); address cumulative environmental and socio-economic impacts; consider options that are only viable at scale (ie across larger geographic areas); establish more extensive and connected ecological habitat corridors; reinstate larger expanses of functional agricultural land; and optimise regional resources (Hattingh, Williams and Corder, 2019).

This paper presents the findings from a multidisciplinary project that identified PMLU options within Queensland's four mining regions: the North-west Minerals Province, North Queensland, Bowen Basin and Clarence-Moreton and Surat Basin (Figure 1). To enable spatial analysis at the appropriate scale and to accommodate geographic diversity, North Queensland was divided into three geographical 'focus areas' – Weipa, Charters Towers and Tablelands. Delineation of the areas captured most of the mines within the wider region as well as regional characteristics and features.

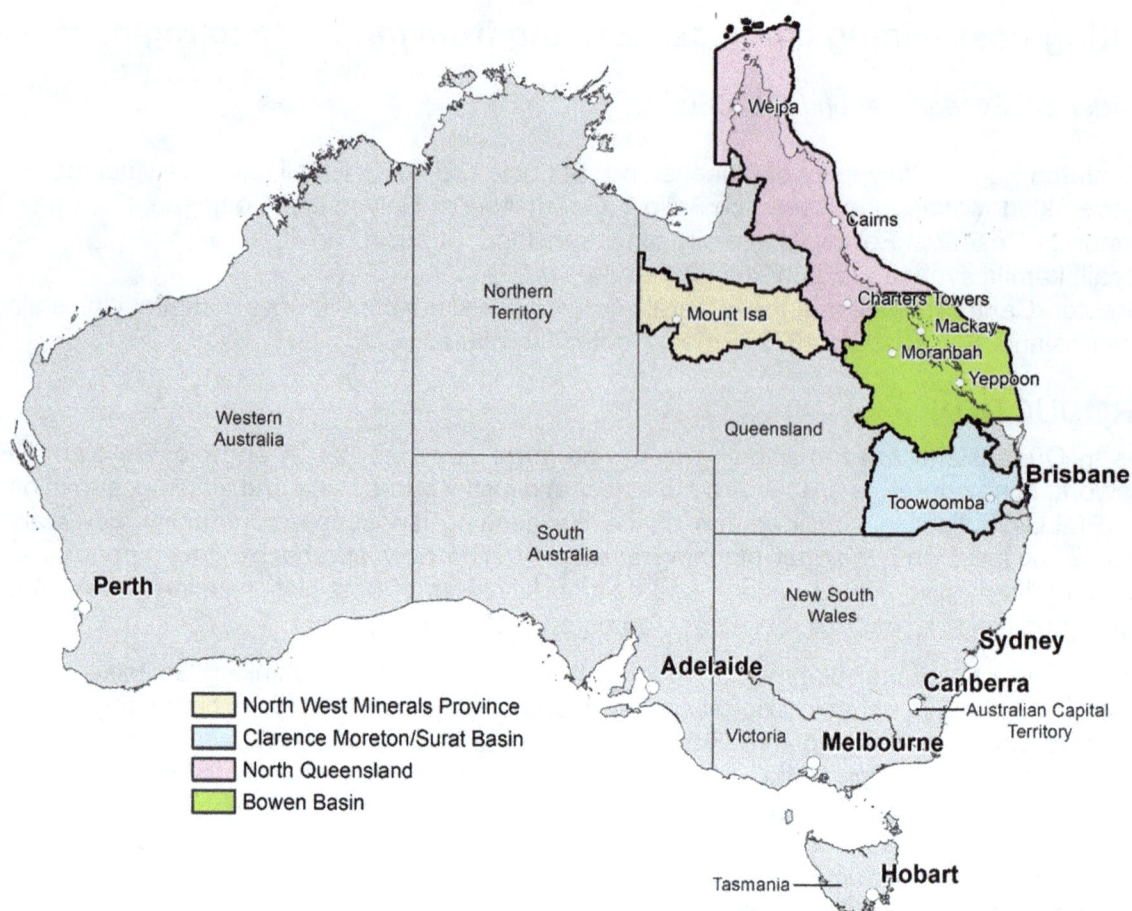

FIG 1 – Australian study regions: North-west Minerals Province, North Queensland, the Bowen Basin and the Clarence-Moreton Surat Basin.

METHOD

For each geographic area, the project team compiled the broad socio-economic and environmental context then produced a shortlist of ten suitable PMLUs based on this contextual data. The shortlist included the conventional choices of native ecosystem corridors and grazing but added wind energy, solar energy, manufacturing hub, tourism, cropping, protected horticulture, intensive livestock and phytomining. To enable comparison of the PMLU options, they were assigned a rating against selected technical indicators, which were then used to allocate a qualitative rating. Options related to agriculture were assessed against access to water or suitable rainfall and suitable soil types; renewable energy options were assessed against climate characteristics (solar radiation/wind speeds) and proximity to a transmission network; and options for biodiversity investments were considered by assessing which areas would derive the most benefits and could reconnect existing corridors. The suitability of the other options (eg intensive livestock, manufacturing, tourism) was based on a qualitative assessment of the presence of required features.

A multicriteria suitability assessment of the PMLU options was then undertaken using the same technical and environmental criteria plus six socio-economic indicators: alignment with regional aspirations, contribution to the regional economy, jobs for regional workers, impact to existing infrastructure and utilities (including demand and maintenance costs), and the skills, education and training of the regional populations and other socio-economic assets to support the PMLU. A five-point Likert scale (very high/very strong to very low) was used to rank the options against the socio-economic indicators. If the PMLU did not meet the region's aspirations nor align with its socio-economic strengths, the suitability was downgraded even if the land use was technically suitable.

The contextual data and suitability assessment outcomes were 'ground-truthed' via two short online surveys and two online scoping discussions with small groups of stakeholders across each study region (ie eight surveys and eight scoping discussions).

RESULTS

The most suitable PMLUs identified for each region are presented in Table 1. These PMLUs were regarded as the most suitable at the time of the assessment. The authors did not seek to exclude nor constrain economic, social and environmental development opportunities that were not identified. At this scale of analysis, small-scale options, such as the repurposing of a mine building for a specific activity, were not assessed. New opportunities driven by global trends, local comparative advantages or economic and social change should be explored over time.

TABLE 1

The most suitable PMLUs identified for each geographic area.

Geographic area	Most suitable PMLUs
North-west Minerals Province	native ecosystem corridors wind energy solar energy
Weipa and Charters Towers – North Queensland	native ecosystem corridors wind energy solar energy tourism protected horticulture
Tablelands – North Queensland	wind energy solar energy tourism grazing
Bowen Basin	native ecosystem corridors solar energy cropping protected horticulture
Clarence-Morton Surat	wind energy tourism grazing protected horticulture

CONCLUSION

Understanding the contextual factors that enable or constrain a region's capacity to transition to a post-mining future is critical for closure planning and decision-making. Many mining regions continue to lag socially and economically decades after mining has ceased (Haggerty et al, 2014). Single PMLUs (eg grazing alone) are unlikely to replace the economic contribution of mining, offer commensurate employment opportunities nor meet regional aspirations, such as having a robust, diverse and sustainable economy or improving the liveability of the region's towns through increased prosperity. Multiple inter-related activities that operate as an integrated system may offer greater socio-economic benefits than single or dual land uses. An integrated system could include, for example, a combination of industrial and agricultural enterprise designed around sustainable water use, renewable energy production, on-site processing and secondary activities, and fieldwork associated with education/training.

Constraints to successful post-mining transition planning abound. Three are highlighted here. First, it is challenging for regional stakeholders to consider transition planning decades in advance as mine closure is intangible at the operational stage of the mining life cycle. It is even more challenging in the current policy environment that is focused on growth, where priority is given to the development of critical minerals operations, hydrogen hubs and renewable energy infrastructure. Second, multi-stakeholder collaboration and decision-making are required to drive post-mining transition planning, however, there is little evidence of this occurring in the state and there is a dearth of regional forums suitable for this purpose. Third, a focus on PMLU limits planning to the mining lease at the expense of the broader regional context. To foster regional resilience to closure-related changes and successful transition to a post-mining future, Queensland needs a strong policy framework around post-mining transition. The framework should integrate PMLU and closure planning into regional planning processes, include collaboration with regional stakeholders and direct resources towards supporting this work.

ACKNOWLEDGEMENTS

The authors would like to thank their colleagues Jo-Anne Everingham, Andrea Arratia-Solar, Pascal Bolz, Mansour Edraki and Peter Erskine for their contribution to the study. They also thank stakeholders from the four study regions who contributed their time and provided valuable insights during the scoping discussions.

The study was commissioned and funded by the Queensland Resources Council and the International Council on Mining and Metals.

REFERENCES

Haggerty, J, Gude, P H, Delorey, M and Rasker, R, 2014. Long-term effects of income specialization in oil and gas extraction: the US West, 1980–2011, *Energy Economics*, 45:186–195.

Hattingh, R, Williams, D J and Corder, G, 2019, September. Applying a regional land use approach to mine closure: opportunities for restoring and regenerating mine-disturbed regional landscapes, in *Mine Closure 2019: Proceedings of the 13th International Conference on Mine Closure*, pp 951–968 (Australian Centre for Geomechanics: Perth).

Effective regulatory frameworks for mine life and beyond

The Mine Land Rehabilitation Authority – reducing risk for Victoria

J Brereton[1] and R Mackay[2]

1. CEO, Mine Land Rehabilitation Authority, Traralgon Vic 3844.
   Email: jennifer.brereton@mineland.vic.gov.au
2. Chair, Mine Land Rehabilitation Authority, Traralgon Vic 3844.
   Email: rae.mackay@mineland.vic.gov.au

INTRODUCTION

On 30 June 2020, the Mine Land Rehabilitation Authority (the Authority) superseded the Latrobe Valley Mine Rehabilitation Commissioner as the independent overseer of Victoria's declared mine rehabilitation. Currently, only the Latrobe Valley's three brown coalmines are declared, with each mine at a different stage of rehabilitation and closure planning: Hazelwood ceased mining in 2017, Yallourn is scheduled to end mining in 2028 and Loy Yang in 2035.

The three declared mines are very large, close together (Figure 1) and present significant environmental risks that must be mitigated during rehabilitation and post closure. In addition to fire risks, the brown coalmines are inherently unstable and require continuous monitoring and management of ground movements and groundwater pressures. Since 2007 several damaging events have occurred, highlighting the difficulties of achieving long-term stability, even with good landform design and rigorous ground control management plans.

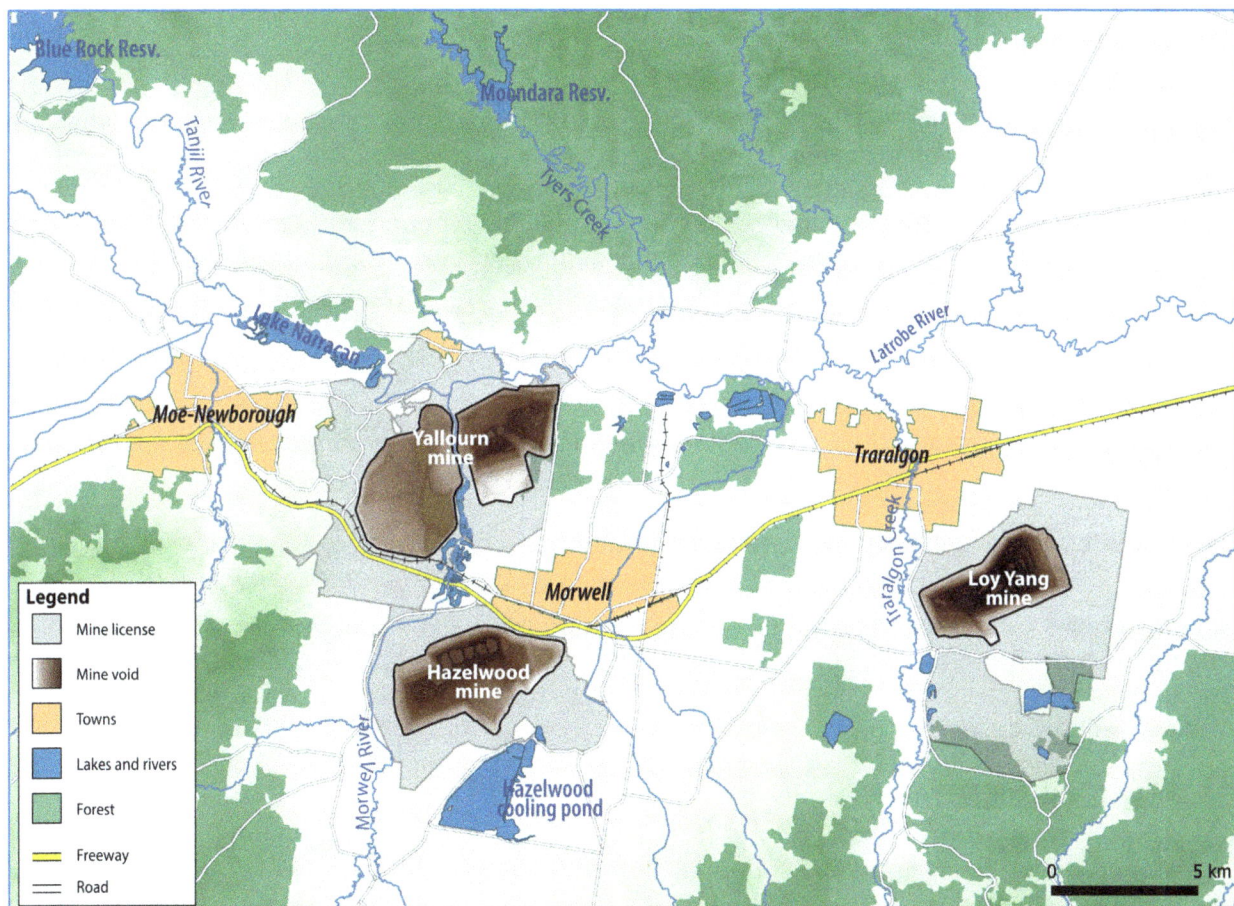

FIG 1 – Plan view of the Latrobe Valley mines.

THE AUTHORITY

Since 2015, the Victorian government has invested in improving the State's regulations governing mine rehabilitation and closure but recognises that environmental risks will remain even with the best rehabilitation plans and practices. These will complicate relinquishment approval and will demand ongoing management by future landowners with oversight by government. The Authority was created by Statute to facilitate both processes.

The Authority not only oversees rehabilitation but is also required to create a register of rehabilitated mine land that defines the controls to be implemented by future landowners post closure. The Authority can own mine land if the magnitude of long-term risks for parcels of land are deemed too high for general ownership. The Authority will assist government to assess the post-closure funds necessary to meet the costs of monitoring registered mine land and to manage the risks of Authority-owned land. A key goal of the Authority is to work with, listen to and involve the community in all matters related to mine rehabilitation and its long-term use.

The Authority's creation and its purpose and functions acknowledge the specific conditions that exist in the Latrobe Valley. The proximity of the mines to each other and to natural and built infrastructure of state significance as well as the fact that the mines are on private land yet require access to state resources to achieve good rehabilitation outcomes set the baseline conditions for integrated management of rehabilitation and closure. The historical significance of brown-coal fired thermal power generation to the economic well-being of the Latrobe Valley, and the social goals of maintaining the economic well-being of impacted communities, at a time of major transition away from thermal power generation underpins the value proposition in ensuring high quality land use outcomes for the mine land released by rehabilitation.

ALTERNATIVE MODELS FOR MANAGING MINE LAND POST CLOSURE

Other models exist for managing mine land post closure, each based on specific regional conditions, but each recognises the requirement for long-term 'in perpetuity' commitments to environmental management. Two examples are in Germany and Canada. Both examples are at different stages of development than Victoria.

In Germany, the RAG-Stiftung (RAG) was created to finance and manage the 'perpetual obligations' resulting from hard coal mining based on commercialising former mining company assets. RAG undertakes capital investments to deliver the returns it needs to manage pit water and groundwater impacted by mining as well as securing the shafts and tunnels of underground mines. RAG also is heavily invested in projects delivering education, science and cultural outputs. If RAG cannot meet long-term rehabilitation obligations, the mining states have provided guarantees. RAG's goal is to avoid this outcome.

In 2005, Saskatchewan, Canada began development of an institutional control framework for the long-term management of decommissioned mine and mill sites on provincial Crown lands. The Institutional Control Program (ICP) was established in 2007. The main components of the ICP are a registry capturing records of closed sites and the required monitoring and maintenance work and the funds to cover: (1) long-term monitoring and maintenance of reclaimed sites in the registry; and (2) an unforeseen events fund to cover costs not foreseen by the monitoring and maintenance program. Sites must meet specific criteria before they can be entered onto the ICP registry.

Other examples can be found in many countries around the world. Work in Australia's other states, notably Queensland and Western Australia, is also developing rapidly to ensure that long-term liabilities are met and that as far as possible the States do not carry the financial burden of managing future environmental risks.

CONCLUDING REMARKS

The Authority provides the platform for ensuring relinquishment can occur, that costs are covered, and community has confidence in long-term access to and management of rehabilitated land. It is one of several models that have been or are being implemented around the world that are all predicated on the specific conditions that prevail in the different regions. The different models present common themes that highlight the complexity of transitioning away from mining to future land uses that deliver positive outcomes for affected communities. Meeting in perpetuity obligations underpin all models.

ACKNOWLEDGEMENTS

We would like to acknowledge the Earth Resources Policy and Programs (within the Department of Energy, Environment and Climate Action) in the Victorian government who undertook the comprehensive overhaul of declared mine rehabilitation for Victoria and who designed and created the Authority's Statute.

Queensland's open cut coalmine void rehabilitation planning practices – challenges and opportunities

M Clay[1] and J Dunlop[2]

1. Senior Environmental Officer, Office of the Queensland Mine Rehabilitation Commissioner, Brisbane Qld 4000. Email: qmrc@qld.gov.au
2. Principal Technical Advisor Resource Rehabilitation, Office of the Queensland Mine Rehabilitation Commissioner, Brisbane Qld 4000. Email: qmrc@qld.gov.au

INTRODUCTION

Thermal and metallurgical coal mining in Queensland has resulted in mines which contain voids due to open cut mining. While there are several proposals for greenfield coalmines or expansions to existing operations, many existing open cut coalmines are reaching maturity and most will leave one or more residual voids in place.

The rehabilitation of mine voids is challenging. Water held in voids will evaporate over time leaving solutes to concentrate. This is likely to lead to poor water quality with limited practical use. The most basic practices to rehabilitate voids have historically involved stabilising the high and low walls, bunding to provide flood protection, and preventing public access.

Regulatory reforms for mine rehabilitation were introduced in Queensland as part of the *Mineral and Energy Resources (Financial Provisioning) Act 2018*, and aimed to strengthen progressive rehabilitation planning, limit risks to the environment and improve outcomes for local and regional communities. Under the reforms, the goal for rehabilitation of voids is to achieve a safe, stable, and non-polluting landform, which can sustain a post-mining land use (PMLU). The reforms also introduced a requirement to describe rehabilitation of mined land in Progressive Rehabilitation and Closure (PRC) plans, replacing the description of such plans in Environmental Authorities (EAs). The reforms are not retrospective and recognise historic approvals for voids, meaning established voids will remain in the landscape as non-use management areas (NUMAs). The objectives of this study were to outline historic void rehabilitation planning practices, describe water quality limitations of Queensland mine voids and identify possible opportunities for improved planning.

PLANNING AND WATER QUALITY OF ESTABLISHED RESIDUAL VOIDS

To understand rehabilitation planning for open cut coalmine voids in Queensland, proposed PMLUs for voids in EAs were examined. Briefly, a data set which collated EAs (Coffey Services Australia Pty Ltd, 2021) was reanalysed, exploring counts and surface areas of open cut coal voids of the Fitzroy Basin and their PMLUs. Both analyses of counts and surface area found that uses for most voids were either not specified in approvals or to be specified in future plans (Figure 1). Broad descriptions for a use such as 'water storage' and 'water filled' were often listed but do not represent a clear future use.

a.
Count of void PMLUs in IESC report

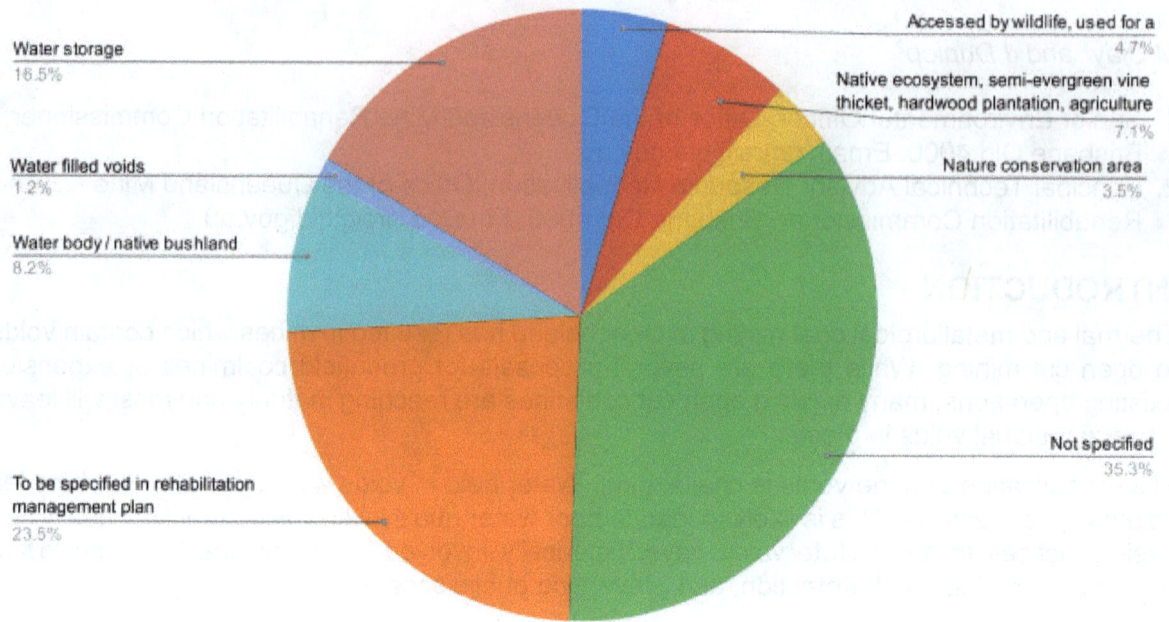

Water storage
16.5%

Water filled voids
1.2%

Water body / native bushland
8.2%

To be specified in rehabilitation
management plan
23.5%

Accessed by wildlife, used for a
4.7%

Native ecosystem, semi-evergreen vine
thicket, hardwood plantation, agriculture
7.1%

Nature conservation area
3.5%

Not specified
35.3%

b.
Planimetric area per PMLU in IESC report

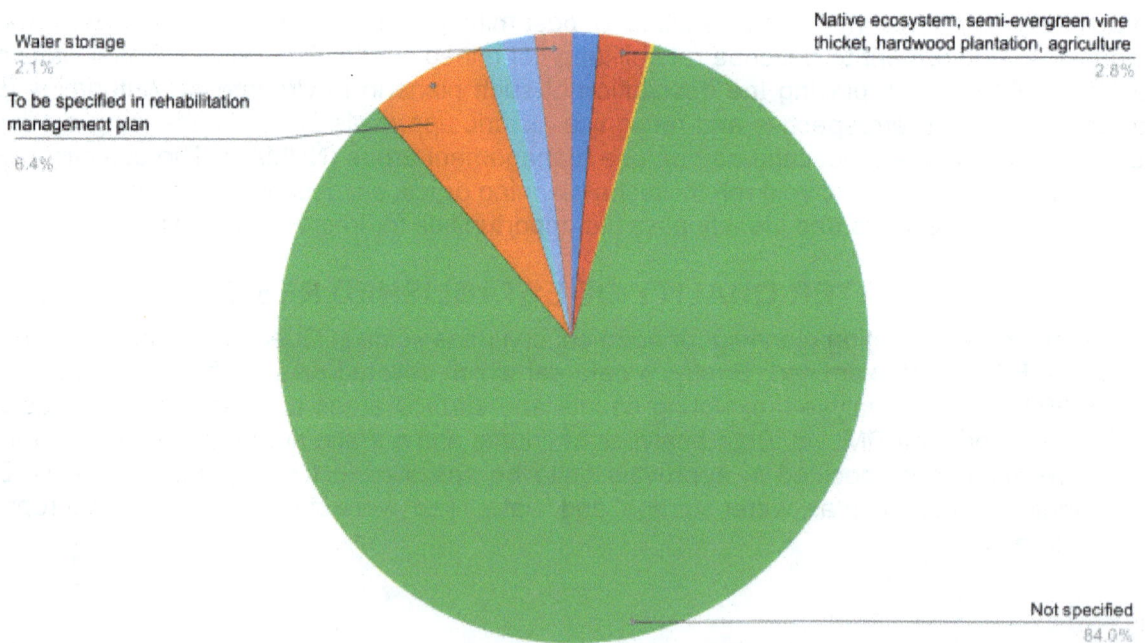

Water storage
2.1%

To be specified in rehabilitation
management plan
6.4%

Native ecosystem, semi-evergreen vine
thicket, hardwood plantation, agriculture
2.8%

Not specified
84.0%

FIG 1 – (a) Counts and (b) surface area of Post Mining Land Uses (PMLUs) assigned to 85 open cut coal voids of the Fitzroy Basin as reported in Coffey Services Australia Pty Ltd (2021).

To examine whether existing water filled voids could host a use, water quality for a suite of variables was compared to a range of use-related guidelines. Publicly available data was retrieved for 12 voids, and compared their water quality across a suite of variables that were highlighted in the literature (Jones *et al*, 2019). Elevated salinity is one of the factors likely to limit PMLUs for many voids within the Fitzroy Basin (Figure 2), with six of the 12 voids containing salinity at levels unsuitable for livestock watering. Accordingly, elevated salinity is likely to limit this PMLU for those voids. Copper, aluminium, and sulfate levels in these voids also showed levels likely to prohibit use. The data set analysed was highly limited by the number of voids and dates of monitoring data (spanning seven years when voids are expected to persist in the environment in perpetuity).

Nonetheless, the monitoring data indicates a trend towards increased salinity, likely due to evapo-concentration, and highlights that it is unlikely that water-filled voids will be able to provide water for agriculture or native aquatic ecosystems without treatment.

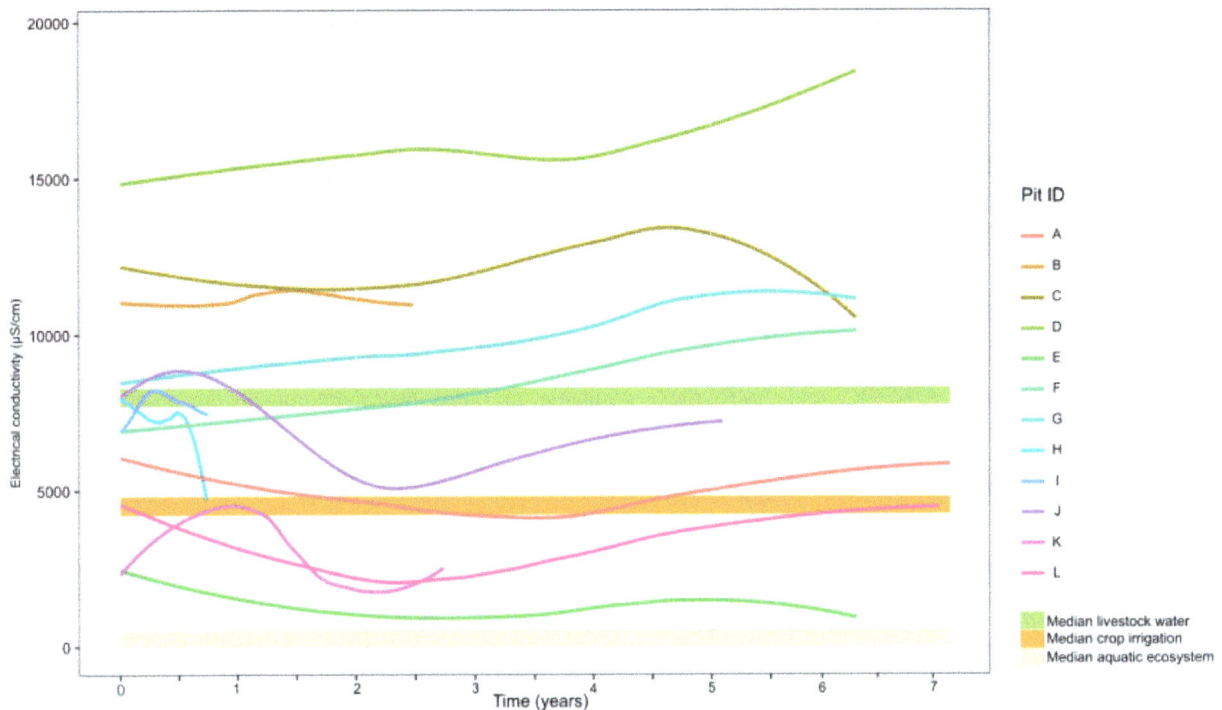

FIG 2 – Electrical conductivity (µS/cm) for 12 open cut coal voids in the Fitzroy Basin compared with use related water quality guidelines (ANZECC and ARMCANZ, 2000; DEHP, 2011; DES, 2017).

CURRENT VOID PLANNING IN PRC PLANS

To understand how mine voids in the Fitzroy Basin are currently being planned for, the counts and surface area of proposed PMLUs for voids in PRC plans were examined. Analysis of PRC plans of nine mines showed that 29 voids are expected to be created during life-of-mine. Twelve of these are planned to be backfilled (to grazing, 11; to native habitat, one), 16 will be NUMAs, and one is proposed to have a PMLU of 'water management'. Although PRC plans have only been completed for a limited number of sites, more are being submitted and approved over time, and will subsequently be analysed for this project.

LOOKING FORWARD – CHALLENGES AND OPPORTUNITIES FOR GREENFIELD OPEN CUT COALMINES

Considering the evidence that water-filled voids in the Fitzroy Basin will likely accumulate water of poor quality, appropriate planning for PMLUs for voids in greenfield sites should clearly show how a use will be achieved. This may include a robust demonstration of water quality, through baseline surveys and modelling, and use viability planning. This may be additional to the *'evidence based comparison and justification for each proposed PMLU against alternative options'* (Queensland Government, 2021, p 21). Alternatives to water-filled void uses involve backfilling a void, which provides flexibility to achieve a range of PMLUs such as grazing or native ecosystems. Backfilling can also help to minimise risks associated with an open water body and avoid leaving an undesirable landform to the local community and future generations.

CONCLUSION

This study outlines a body of evidence which shows that many open cut coalmines in Queensland have approved residual voids. These voids are likely to remain in the landscape and accumulate increasingly poor-quality water. The data analysed here shows that it will be difficult or costly to retrofit these voids for uses that are compliant in terms of water quality and would potentially involve ongoing water treatment. Our analysis of PRC plans shows increased clarity in planning for voids, although NUMAs have been proposed within those documents. Water-filled voids at greenfield sites and new expansions require careful consideration given water quality limitations of extant residual voids. Resultingly, there is an opportunity for good planning that will not compromise regional water quality, will strive for leading practice rehabilitation and will benefit the local environment and communities.

ACKNOWLEDGEMENTS

We would like to thank the OQMRC team for their support of the project and their reviews of drafts, notably James Purtill, Louisa Nicolson, and Kate Baker. We would also like to thank Dr Lucy Reading and Dr Robynne Crystal for their guidance during the project.

REFERENCES

Australia and New Zealand Environment and Conservation Council Agriculture and Resource Management Council of Australia (ANZECC and AMRCANZ), 2000. *Australian and New Zealand Guidelines for Fresh and Marine Water Quality: Volume 1*, ANZECC and AMRCANZ. Available at: https://www.waterquality.gov.au/anz-guidelines/resources/previous-guidelines/anzecc-armcanz-2000 (Accessed: 26 April 2022).

Coffey Services Australia Pty Ltd, 2021. Scoping study – Coal mine voids, Queensland: A report commissioned for the Independent Expert Scientific Committee on Coal Seam Gas and Large Coal Mining Development, Independent Expert Scientific Committee on Coal Seam Gas and Large Coal Mining Development (IESC).

Department of Environment and Heritage Protection (DEHP), 2011. *Environmental Protection (Water) Policy 2009 Fitzroy River Sub-basin Environmental Values and Water Quality Objectives Basin No. 130 (part), including all waters of the Fitzroy River Sub-basin,* Department of Environment and Heritage Protection.

Department of Environment and Science (DES), 2017. *Guideline: Model mining conditions,* Department of Environment and Science. Available at: https://environment.des.qld.gov.au/__data/assets/pdf_file/0033/88926/rs-gl-model-mining-conditions.pdf (Accessed: 25 May 2022).

Jones, C E, Vicente-Beckett, V and Chapman, J, 2019. Coal mine-affected water releases, turbidity and metal concentrations in the Fitzroy River Basin, Queensland, Australia, *Environmental Earth Sciences*, 78(24):1–16, doi:10.1007/s12665–019–8734-x.

Queensland Government, 2021. *Guideline – Progressive rehabilitation and closure plans (PRC plans)*, Department of Environment and Science, Queensland Government. Available at: https://environment.des.qld.gov.au/__data/assets/pdf_file/0026/95444/rs-gl-prc-plan.pdf.

The role of the regulator in promoting a highly reliable mining industry

S Johnston[1], S Stevens[2] and L Howe[3]

1. Program Leader, Leadership and Organisational Improvement, Sustainable Minerals Institute, The University of Queensland, St Lucia, Qld, 4072. Email: susan.johnston@uq.edu.au
2. Manager Environmental Leadership and Performance, Centre for Water in the Minerals Industry, Sustainable Minerals Institute, The University of Queensland, St Lucia, Qld, 4072. Email: shona.stevens@uq.edu.au
3. PhD Candidate, Centre for Water in the Minerals Industry, Sustainable Minerals Institute, The University of Queensland, St Lucia, Qld, 4072. Email: layla.howe@uq.edu.au

INTRODUCTION

Regulators have an important role to play in fostering and enabling reliable, effective, environmental management. Given this, it is perhaps surprising that there has been relatively little debate on exactly how regulators can enhance their own approaches, in order to promote improved outcomes overall. The authors argue that applying research on High Reliability Organisations, as well as the University of Pennsylvania's Penn Program Framework for Regulatory Excellence, to regulators has the potential to facilitate positive environmental results across the mine life cycle.

HIGH RELIABILITY ORGANISATIONAL THINKING

High Reliability Organisational (HRO) thinking has traditionally been seen as an approach that aims to improve health and safety within organisations. In fact, HRO thinking can be used to improve the reliability and effectiveness of any aspect of any organisation's activities (Sutcliffe, 2018). The research tells us that organisations that exhibit HRO characteristics routinely deliver on their goals (Weick and Sutcliffe, 2015; Cantu et al, 2020; Haslam et al, 2022; Dwyer, Karanikas and Sav, 2023). In the environmental regulator space, those goals typically include an intention to facilitate improved environmental management in regulated industries. In the authors' view, regulators who adopt and embody HRO thinking will be far better placed to influence in this regard.

HIGH RELIABILITY ORGANISATION ATTRIBUTES

Becoming a more highly reliable organisation requires an honest assessment of current organisational strengths and weaknesses. It requires an evaluation of the extent to which an organisation aligns with the core HRO characteristics. More highly reliable organisations have:

- a distinct and shared sense of purpose

- a sensitivity to, and a focus on, identifying and addressing organisational issues such as silos and poor communication

- a reluctance to simplify complex issues

- a willingness to defer to expertise no matter where that expertise originates

- a preoccupation with anticipating and avoiding failure

- a commitment to resilience and to ensuring that adequate resources are in place to both prevent, and respond, to failure when it does occur.

If we were to consider an example, the multi-faceted planning aspects for closure of a mine site often needs engagement with a range of distinct regulatory departments with different legislative powers relating to various aspects like water, land or biodiversity, to list a few. Application of HRO principles could involve greater value being placed by regulators on removing their organisational barriers operating as a comprehensive regulatory system that minimise actions occurring in silos. Closure strategies could then more easily align with a regional vision of post-mining future.

Researchers at The University of Queensland have developed a diagnostic methodology that can be used to assess HRO alignment as a first step towards organisational improvement. Applying that methodology to regulators would be a novel approach to enhancing their ability to positively impact environmental management. It would also be an approach grounded in more than three decades of research and experience elsewhere.

REGULATORY EXCELLENCE – THE PENN PROGRAM REVIEW

In contrast, the Penn Program Framework for Regulatory Excellence (PPR) actually derives from a 2014 initiative by Canada's Alberta Energy Regulator (AER). The AER, which had accountability for assessing environmental, and other, performance across the Alberta resources sector, sought to understand what constituted a 'best in class' Regulator (Coglianese, 2015). Multiple workshops, interviews, and research papers followed, leading, ultimately, to development of a regulatory framework that could be adopted by all regulators, everywhere.

The PPR is based on the view that excellent regulators have clarity about exactly how they can enable and influence appropriate behaviours on the part of those they regulate. Excellent regulators go well beyond enforcement (though this is important), to understand, deeply, what actions they need to take to constructively challenge, and support regulated companies. Excellent regulators embody three core attributes, (namely 'empathic engagement', 'stellar competence', and 'utmost integrity'), and are committed to 'nine tenets of regulatory excellence' (Coglianese, 2015). While some of these attributes, such as the need for technical competency, are unsurprising, others challenge traditional views of the scope of a regulator's role. One of the PPR tenets, for example, suggests that excellent regulators will 'initiate' 'productive public dialogue on issues relevant to the regulator's mission' with the idea being that regulated entities, and other stakeholders, help to refine the regulator's mission and practices on an ongoing basis (Coglianese, 2015).

As an example, from the perspective of mine closure, this principle illustrates the importance of the role regulators can play to facilitate and support constructive dialogue with external stakeholders interested in, or affected by, the closure planning process at a local and potentially regional scale. In other words, regulators applying this tenet would aim to gather informative commentary relating to closure planning decisions from all affected external stakeholders.

Another tenet emphasises the importance of regulators avoiding the trap of seeing industry outcomes as direct measures of regulator performance, without first establishing the 'causal connection' between the two (Coglianese, 2015). The PPR also takes a broad view of regulator competence. Regulators need to have the 'right attitudes', including 'humility and empathy' as well as the necessary technical skills (Coglianese, 2015).

Since the PPR was proclaimed, regulators around the world have made varying levels of commitment to implementation of the framework. The Queensland Department of Regional Development, Manufacturing and Water, for example, has instituted a 'performance excellence framework' which draws on some components of the PPR. However, the full suite of PPR characteristics has yet to be applied to regulators charged with overseeing environmental performance in the Australian resources sector.

TIME FOR A DEEPER CONVERSATION

As the Australian mining industry looks to overcome a myriad of environmental challenges, and as regulation is reformed to promote environmental improvement, it is important that the regulators themselves continually seeking to achieve 'best in class' outcomes within their span of control. The authors believe that the combination of application of High Reliability Organisational thinking, and the Penn Program Framework for Regulatory Excellence, to mining industry regulators has the potential to ensure that those regulators are significant enablers of industry transformation. The time is right for a deeper conversation, involving multiple stakeholders, on what a good environmental regulator in the Australian mining industry context looks like.

REFERENCES

Cantu, J, Tolk, J, Fritts, S and Gharehyakheh, A, 2020. High Reliability Organization (HRO) systematic literature review: Discovery of culture as a foundational hallmark, *Journal of Contingencies Crisis Manag*, 28:399–410. https://doi.org/10.1111/1468–5973.12293

Coglianese, C, 2015. *Listening Learning Leading: A Framework for Regulatory Excellence*, Penn Program on Regulation, Penn Program on Regulation: Philadelphia, U S.

Dwyer, J, Karanikas, N and Sav, A, 2023. Scoping review of peer-reviewed empirical studies on implementing high reliability organisation theory, *Safety Sci,* 164:106178. https://doi.org/10.1016/j.ssci.2023.106178

Haslam, S A, Jetten, J, Maskor, M, McMillan, B, Bentley, S V, Steffens, N K and Johnston, S, 2022. Developing high-reliability organisations: A social identity model, *Saf Sci*, 153:105814. https://doi.org/10.1016/j.ssci.2022.105814

Sutcliffe, K, 2018. Mindful Organising, in: *Organizing for Reliability: A Guide for Research and Practice* (eds: R Ramanujam and K Roberts), pp 61–89 (Stanford University Press: California).

Weick, K and Sutcliffe, K, 2015. *Managing the Unexpected: Sustained Performance in a Complex World* (John Wiley & Sons).

Embedding the circular economy into life-of-mine planning

Economic analyses (EA) and life cycle assessment (LCA) on repurposing of mine waste via geopolymerisation technology

S Amari[1], M Darestani[2], H M A Ilyas[3] and M Yahyaei[4]

1. Research Fellow, Julius Kruttschnitt Mineral Research Centre, Sustainable Minerals Institute (SMI), The University of Queensland, St Lucia Qld 4068. Email: s.amari@uq.edu.au
2. Senior Lecturer, School of Mechanical Engineering, Western Sydney University, Parramatta NSW 2150. Email: m.darestani@westernsydney.edu.au
3. Postdoctoral Research Fellow, Julius Kruttschnitt Mineral Research Centre, Sustainable Minerals Institute (SMI), The University of Queensland, St Lucia Qld 4068. Email: h.ilyas@uq.edu.au
4. Professor, Julius Kruttschnitt Mineral Research Centre, Sustainable Minerals Institute, The University of Queensland, St Lucia Qld 4072. Email: m.yahyaei@uq.edu.au

INTRODUCTION

Tailings are generally deposited on or near to the mine sites, occupying the site area, and being an economic burden for mining companies. Tailings can cause several environmental impacts, such as leaking toxic substances and contaminating underground water, emitting greenhouse gases, and damaging natural bodies and wildlife. Geopolymerisation of mine waste effectively transforms mine waste into more valuable, environmentally friendly products. Geopolymers are inorganic aluminosilicate materials that are widely applied in the construction industry (Araya, Kraslawski and Cisternas, 2020). The high performance and the low-CO_2 emission of geopolymers make this type of material potentially more desirable than ordinary cement, given that manufacturing one ton of traditional cement generates approximately 0.8 tons of CO_2. This process is seen as a promising method for repurposing tailings and reducing the environmental impact of mining waste. However, the environmental impact of geopolymer production is not sufficiently comprehended (Mazzinghy et al, 2022). Moreover, geopolymerisation of mine waste can offer several economic benefits, including reduced waste management costs (Amari et al, 2019; He et al, 2022), reduced raw material costs, and potentially lower production costs. However, there are several challenges that need to be addressed to ensure the economic feasibility of the process.

This study aims to demonstrate the potential environmental impacts of geopolymer concrete produced using mine wastes and industrial by-products such as blast furnace slag (BFS). For this, a comparative cradle-to-gate life cycle assessment (LCA) methodology was conducted on different geopolymer mix designs, and the greenhouse gas (GHG) emissions results were compared to the ordinary cement. LCA can be used to evaluate the environmental impact of geopolymerisation of mine wastes in various ways. One important impact category is GHG gas emissions, which contribute to climate change and is the focus of this study. Another aim of this study is to evaluate the economic feasibility of the geopolymerisation of mine wastes via economic analysis (EA) to identify the key economic factors and trade-offs associated with repurposing mine wastes through geopolymerisation.

MATERIALS AND METHODS

Materials

Experiments involved the use of waste zeolite from Australia with particles size of below 54 μm. Ground BFS with an average particles size of <35 μm was supplied from Independent Cement and Lime Pty. Ltd. The chemical composition of the mined zeolite and BFS as determined by X-ray fluorescence (XRF) analyses is given in Table 1. Sodium hydroxide solution with a concentration of 40% w/w was obtained from Chem-Supply Pty Ltd and diluted to a concentration of 5 M. Sodium silicate solution (grade D) with a mass ratio of $SiO_2/Na_2O = 1.95–2.05$ was purchased from PQ Australia Pty Ltd and used with no change.

TABLE 1

Chemical composition of mined zeolite and BFS using X-ray fluorescence analyses.

	Component oxide (wt.%)											
	SiO_2	Al_2O_3	CaO	K_2O	Fe_2O_3	MgO	Na_2O	TiO_2	P_2O_5	SO_3	MnO	LOI*
Waste zeolite	67.49	12.18	3.04	1.83	1.43	0.95	0.94	0.21	0.03	0.01	0.06	11.83
BFS	32.35	13.32	41.81	0.32	0.29	5.31	0.21	0.55	0.02	2.73	0.21	1.51

LOI: Loss on ignition at 1050°C.

Geopolymer preparation

The experiment incorporated mined zeolite, 5 M sodium hydroxide solution, and sodium silicate (grade D with a mass ratio of SiO_2/Na_2O = 1.95–2.05), which were combined to create a uniform slurry. The BFS was introduced in varying proportions of 10, 20, or 30% wt to the mixture, alongside the necessary amount of water, ensuring a solid/liquid mass ratio of 1.3 and a sodium hydroxide/sodium silicate mass ratio of 3.5 and water/solid mass ratio: 7.7 per cent. This was done strategically to prevent premature hardening due to the quick setting nature of BFS. The mixture was subsequently moulded into cylinders, subjected to curing at room temperature, 40°C, and 60°C for 24 hours, and sealed to avoid over-drying. To maintain experimental precision, the process was repeated in triplicate, with average data reported.

Measuring GHG emissions through life cycle assessment

The life cycle inventory (LCI) of geopolymer production was developed at a laboratory scale to evaluate the environmental impacts associated with the production process (Figure 1). The LCI includes inputs and outputs associated with the geopolymerisation process, including raw material extraction, transportation, manufacturing, and use (Munir *et al*, 2023; Zhang *et al*, 2023). For this study, BFS, chemical solutions, mine waste, and the energy and water inputs were considered for the process. However, this study aims to evaluate the potential environmental impacts at an industrial scale because the environmental impacts can change significantly due to differences in energy and resource requirements, thus the functional unit for the analysis is considered 1 m^3 concrete. This study follows relevant standards and guidelines such as ISO 14040 and ISO 14044 and a specific guideline for conducting LCAs of construction materials ie EN 15804. Part of data was obtained from literature and testing. Other parts were obtained from databases such as Ecoinvent and USLCI.

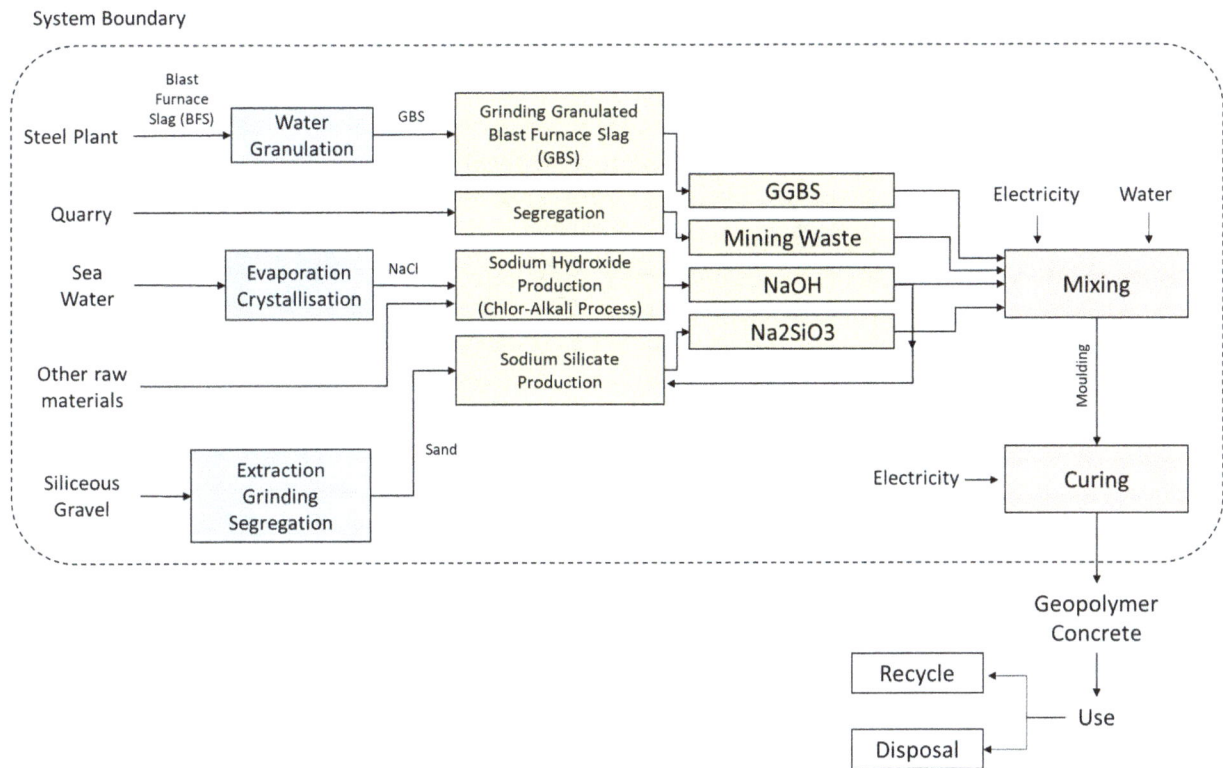

FIG 1 – Life cycle assessment of geopolymerisation process of mine waste.

Economic analyses

For the economic evaluation, economic analysis method was used to assess the economic feasibility of a process or technology by analysing the costs and benefits over its entire life cycle. The EA involves estimating capital costs, estimating operating costs, conducting a cost-benefit analysis, to evaluate the impact of key assumptions and uncertainties on the economic viability of the geopolymer production process using aluminosilicate wastes (Gao *et al*, 2023). In this study, the costs for equipment and infrastructure for batching, mixing, and placing of the concrete have been considered same for both geopolymer and conventional concrete production. This assumption provides a baseline for comparison and simplifies the economic analysis.

Results and discussion

The comparative analysis of the GHG emissions between geopolymer concrete and conventional cement concrete provides us with clear evidence of the environmental advantage of geopolymerisation technology. Our analysis showed that the emissions ranged between 205 and 240 kg CO_2-e/m^3 for geopolymer concrete, which is significantly lower than the 354 kg CO_2-e/m^3 emissions from conventional concrete production. This difference of approximately 30 per cent to 40 per cent is substantial and reaffirms the environmental efficacy of the geopolymerisation process. However, a crucial factor to consider in this comparison is the mechanical strength of the resulting material. The maximum strength achieved by our geopolymer concrete samples was 12 MPa after 24 hr, significantly lower than the 30 MPa commonly achieved by traditional cement concrete. This disparity in early strength might limit the potential applications of geopolymer concrete or require additional materials or processes to enhance its mechanical strength.

In comparing costs, geopolymer concrete, ranging from $430 to $445/m^3 (Figure 2), is more expensive than traditional concrete at $283/m^3. This difference is primarily due to the chemicals used in geopolymer production. However, these costs might offset against the environmental benefits of geopolymer concrete. It's important to note that these costs could vary depending on location and market conditions.

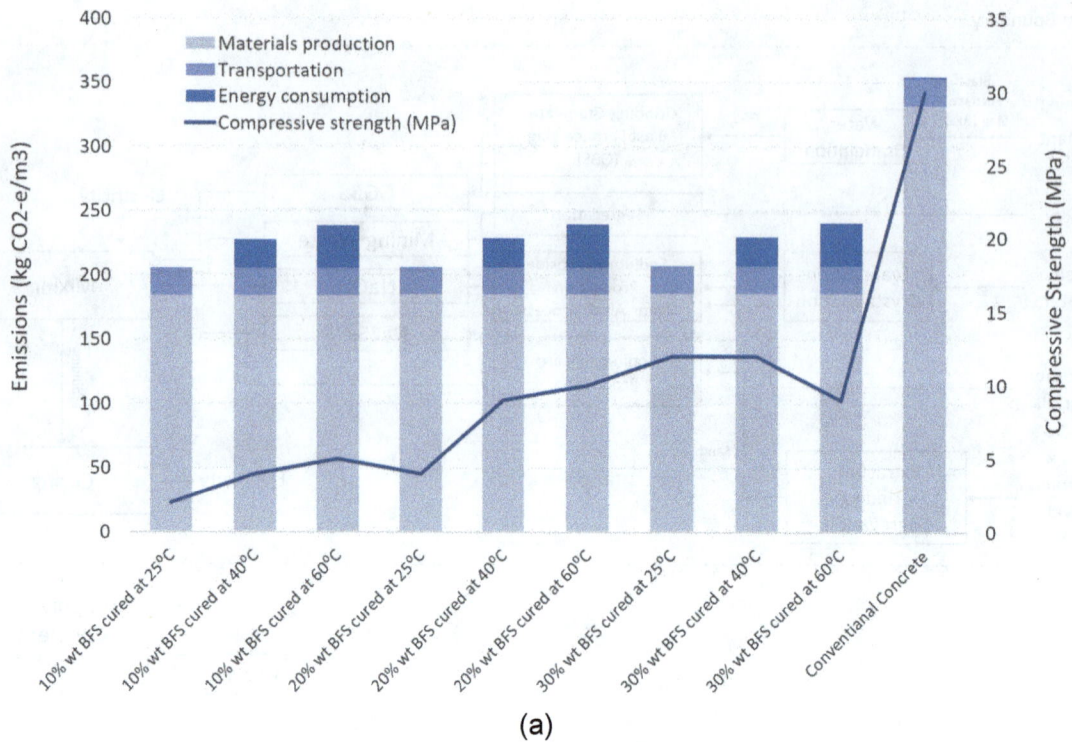

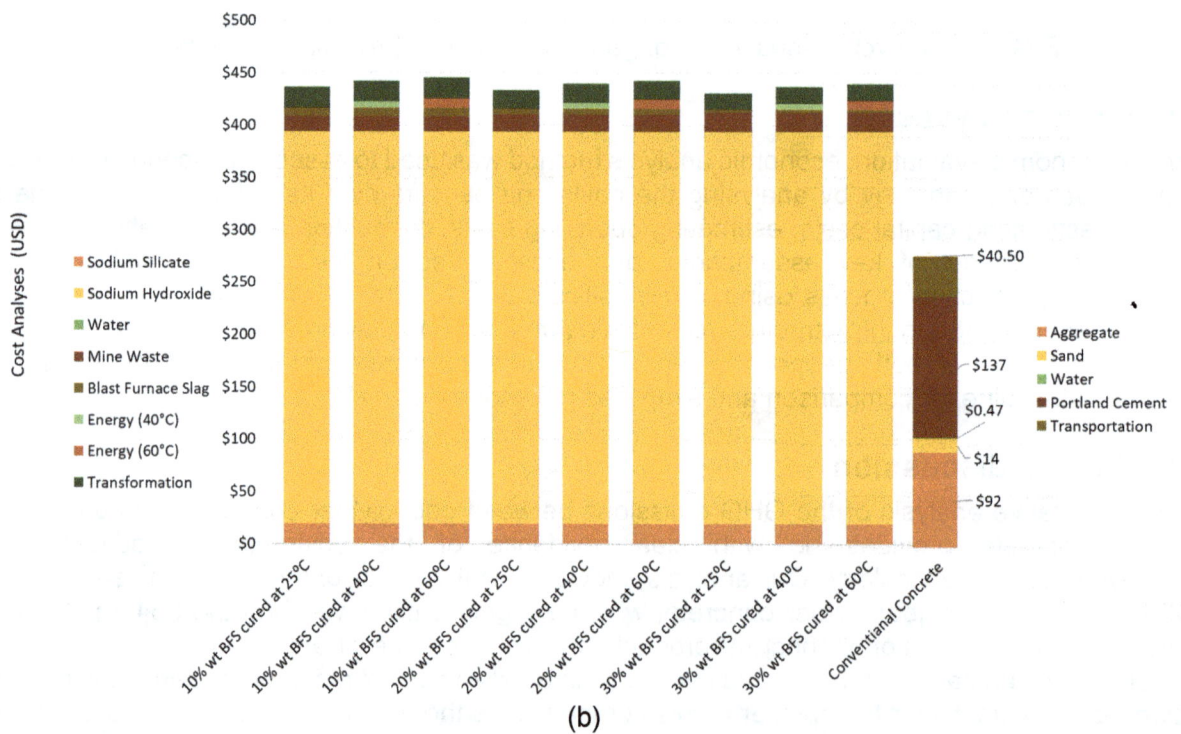

FIG 2 – (a) Emissions from geopolymerisation process of mine waste; (b) Economic assessment of geopolymer production using mine wastes.

REFERENCES

Amari, S, Darestani, M, Millar, G J, Rintoul, L and Samali, B, 2019. Microchemistry and microstructure of sustainable mined zeolite-geopolymer, Journal of Cleaner Production, 234:1165–1177, https://doi.org/10.1016/j.jclepro.2019.06.237

Araya, N, Kraslawski, A and Cisternas, L A, 2020. Towards mine tailings valorization: Recovery of critical materials from Chilean mine tailings, *Journal of Cleaner Production*, 263:121555, https://doi.org/10.1016/j.jclepro.2020.121555

Gao, Z, Li, Y, Qian, H and Wei, M, 2023. Environmental, economic, and social sustainability assessment: A case of using contaminated tailings stabilized by waste-based geopolymer as road base, Science of The Total Environment, 888:164092, https://doi.org/10.1016/j.scitotenv.2023.164092

He, X, Yuhua, Z, Qaidi, S, Isleem, H F, Zaid, O, Althoey, F and Ahmad, J, 2022. Mine tailings-based geopolymers: A comprehensive review, Ceramics International, 48(17):24192–24212, https://doi.org/10.1016/j.ceramint.2022.05.345.

Mazzinghy, D B, Figueiredo, R A M, Parbhakar-Fox, A, Yahyaei, M, Vaughan, J and Powell, M S, 2022. Trialling one-part geopolymer production including iron ore tailings as fillers, International Journal of Mining, Reclamation and Environment, 36(5), https://doi.org/10.1080/17480930.2022.2047271

Munir, Q, Abdulkareem, M, Horttanainen, M and Kärki, T, 2023. A comparative cradle-to-gate life cycle assessment of geopolymer concrete produced from industrial side streams in comparison with traditional concrete, Science of The Total Environment, 865:161230, https://doi.org/10.1016/j.scitotenv.2022.161230

Zhang, J, Fernando, S, Law, D W, Gunasekara, C, Setunge, S, Sandanayake, M and Zhang, G, 2023. Life Cycle Assessment for Geopolymer Concrete Bricks Using Brown Coal Fly Ash, Sustainability, 15(9):7718, https://doi.org/10.3390/su15097718

Exploring the applicability of responsible research and innovation for mine transitions

A Lee Zhi Yi[1], S Worden[2], A Littleboy[3], G Corder[4] and E Sellers[5]

1. PhD Candidate, Sustainable Minerals Institute, University of Queensland, St Lucia Qld 4067. Email: a.leezhiyi@uq.edu.au
2. Research Fellow, Sustainable Minerals Institute, University of Queensland, St Lucia Qld 4067. Email: s.worden@uq.edu.au
3. Professorial Research Fellow, Sustainable Minerals Institute, University of Queensland, St Lucia Qld 4067. Email: a.littleboy@uq.edu.au
4. Honorary Associate Professor, Sustainable Minerals Institute, University of Queensland, St Lucia Qld 4067. Email: g.corder@uq.edu.au
5. Research Director, Hard Rock Mining, CSIRO, Pullenvale Qld 4069. Email: ewan.sellers@csiro.au

MINE TRANSITIONS AND MINE LIFE EXTENSION TECHNOLOGIES (MLET)

The global shift towards decarbonisation is driving competition for the extraction of critical minerals such as nickel, lithium, and copper. To meet this demand, new mines are being developed while mines approaching closure are being reimagined as sites of potential resources. This reimagining challenges how mine closure is being conceptualised. In the conventional case, when a mine reaches the end of its economic life, it is closed or put into care and maintenance until it transitions to the next land use, such as agriculture, native ecosystem corridors, solar farms or integrated eco – precincts (Murphy *et al*, 2019; Holcombe, 2020) (the authors acknowledge that unplanned mine closures have left a devastating legacy of contamination, safety and cultural heritage impacts). Extracting minerals from closed or closing mines will extend mine life and, therefore, delay relinquishment and transition.

In this context, innovative technologies are needed to make mining economically viable. In the first year of her doctoral research on improving stakeholder engagement in mine transitions, Lee Zhi Yi has coined the term 'mine life extension technologies' (MLETs) to describe a class of technological solutions that optimise extraction of resources after a mine reaches the end of its economic life. Examples of MLETs include agromining (van der Ent *et al*, 2015), *in situ* recovery (Robinson and Kuhar, 2018), and reprocessing or reuse of tailings (Gou, Zhou and Then, 2019).

MLET extraction offers several benefits compared with conventional mining processes, such as waste remediation, value recovery of minerals, a smaller footprint and less environmental impacts (Sellers, Picorelli and Salmi, 2022). What requires further investigation is the consequences of MLET implementation to host communities or community perspectives towards the proposed technology and subsequent delays in transition.

COMMUNITY ENGAGEMENT

The importance of engaging with host communities about mining projects and technology implementation is well documented in the grey and academic literature (Everingham, 2007; Moffat and Zhang, 2014; Minerals Council of Australia, 2015; OECD, 2017). Engagement is the process of building ongoing relationships and trust with host communities and other stakeholders, which should start early in the mining life cycle and extend throughout the mine life (ICMM, 2019). Wilson *et al* (2017) note that engagement can range from communities being 'informed' and 'consulted with' to their active involvement in decision-making about matters that affect them. Increasingly, the term 'meaningful' is being applied to community engagement to reflect an approach that is ongoing, two-way, conducted in good faith and responsive (OECD, 2017). Meaningful engagement can help tenement holders understand the social risks their developments may pose to host communities and provide opportunities to work together to identify solutions.

The scholarly literature abounds with articles that position mining as a socially risky proposition, from risks to human rights (Kemp and Vanclay, 2013), to livelihoods (Adam, Owen and Kemp, 2015), cultural heritage (O'Faircheallaigh, 2008), equitable distribution of benefits (Bebbington and Bury, 2009) and connection to place (Svobodova, Plieninger and Sklenicka, 2023). Kemp, Worden and Owen (2016) contend that the mining industry's current approach to social risk conflates risk to people and risk to the business, thereby obfuscating how and where both types of risk interact. The authors argue that failure to address social risks can rebound as business risks and they point to Franks *et al*'s (2014) cost of conflict study as an example of the rebound concept in action. The study shows how conflict translates social and environmental risk into costs to the business.

In the MLET extraction context, engagement with community stakeholders often occurs after the tenement holder has selected the technology and is working with the technology provider to progress with on-site implementation. At this stage of the process, there is little scope for the community to share its views and concerns about the MLET or for social risks to be identified and addressed. The failure to engage with the community ahead of implementation could result in business risks such as curtailed implementation or costly retrospective adaptation if the technology is not socially acceptable. Engaging with community stakeholders earlier in the MLET implementation process is a means of identifying and, potentially, addressing social and business risks.

RESPONSIBLE RESEARCH AND INNOVATION TO IMPROVE MLET IMPLEMENTATION

Responsible Research and Innovation (RRI) is a technology assessment process that could improve MLET application, as it encourages early and meaningful engagement between technology providers and community stakeholders (von Schomberg, 2014). RRI processes are beneficial to technology providers and community stakeholders. RRI ideation encourages potential implications and societal expectations of technologies to be considered, allowing MLET developers an opportunity to adapt to community needs and minimise business risks. Meanwhile, through RRI activities such as focus groups and formal evaluations (Long *et al*, 2020), community stakeholders are encouraged to share their views during the technology development process, allowing for an opportunity to discuss and jointly address social risks. Co-development of solutions may, in principle, lead to innovations that are more effective and trusted (Long *et al*, 2020), which would allow for more successful MLET implementation.

While much work has been performed in understanding the social implications of RRI in the policy domain, its feasibility and operationalisation in the commercial domain – specifically for mine transitions and MLET implementation – has yet to be assessed. Aside from peripheral research examining RRI and emerging technologies for climate engineering (Low and Buck, 2020) and a workshop on RRI implementation in the Chilean mining industry (Tavlaki, 2015), there is a scarcity of information on the subject. Further research is required to explore the applicability of RRI in this context.

Applying RRI to MLET implementation promotes early engagement and co-development of solutions and thereby provides an opportunity to reshape the solution-oriented paradigm of the mining industry, and open conversations between technology providers and the community. Given the potential for both to result in better mine transition outcomes, Lee Zhi Yi's doctoral research is examining the applicability of RRI to mine transitions and MLET implementation.

ACKNOWLEDGEMENTS

This research was performed with the generous support of the CRC Transition in Mining Economies (CRC TiME) Top-up Scholarship and the University of Queensland Research Training Program Scholarship.

REFERENCE LIST

Adam, A B, Owen, J R and Kemp, D, 2015. Households, livelihoods and mining-induced displacement and resettlement, *The Extractive Industries and Society*, 2(3):581–589.

Bebbington, A J and Bury, J T, 2009. Institutional challenges for mining and sustainability in Peru, *Proceedings of the National Academy of Sciences*, 106(41):17296–17301.

Everingham, J-A, 2007. Towards Social Sustainability of Mining, *Greener Management International*, 2007(57):91–103.

Franks, D, Davis, R, Bebbington, A J, Ali, S H, Kemp, D and Scurrah, M, 2014. Conflict translates environmental and social risk into business costs, in *Proceedings of the National Academy of Sciences*, 111(21):7576–7581.

Gou, M, Zhou, L and Then, N W Y, 2019. Utilization of tailings in cement and concrete: A review, *Science and Engineering of Composite Materials*, 26(1):449–464.

Holcombe, S, 2020. *Woodlawn mine site repurposing: Success factors, enablers and challenges,* Brisbane: Centre for Social Responsibility in Mining, The University of Queensland, The Social Aspects of Mine Closure Research Consortium.

International Council on Mining and Metals (ICMM), 2019. *Integrated Mine Closure — Good Practice Guide,* International Council on Mining and Metals. Available from: <https://guidance.miningwithprinciples.com/integrated-mine-closure-good-practice-guide/>

Kemp, D and Vanclay, F, 2013. Human rights and impact assessment: clarifying the connections in practice, *Impact Assessment and Project Appraisal*, 31(2):86–96.

Kemp, D, Worden, S and Owen, J, 2016. Differentiated social risk: Rebound dynamics and sustainability performance in mining, *Resources Policy*, 50:19–26.

Long, T B, Blok, V, Dorrestijn, S and MacNaughten, P, 2020. The design and testing of a tool for developing responsible innovation in start-up enterprises, *Journal of Responsible Innovation*, 7(1):45–75.

Low, S and Buck, H J, 2020. The practice of responsible research and innovation in 'climate engineering'. *WIREs Climate Change*, 11(3):e644.

Minerals Council of Australia, 2015. Enduring Value Principles – The Australian Minerals Industry Framework for Sustainable Development.

Moffat, K and Zhang, A, 2014. The paths to social licence to operate: An integrative model explaining community acceptance of mining, *Resources Policy*, 39:61–70.

Murphy, D, Fromm, J, Bairstow, R and Meunier, D, 2019. A repurposing framework for alignment of regional development and mine closure, in *Mine Closure 2019: Proceedings of the 13th International Conference on Mine Closure* (eds: A B Fourie and M Tibbett), pp 789–802 (Australian Centre for Geomechanics: Perth). https://doi.org/10.36487/ACG_rep/1915_64_Murphy

O'Faircheallaigh, C, 2008. Negotiating Cultural Heritage? Aboriginal–Mining Company Agreements in Australia, *Development and Change*, 39(1):25–51.

Organisation for Economic Co-operation and Development (OECD), 2017. *OECD Due Diligence Guidance for Meaningful Stakeholder Engagement in the Extractive Sector,* Organisation for Economic Co-operation and Development.

Robinson, D and Kuhar, L, 2018. Extending Mine Life Through Application of an In-Situ Recovery Approach.

Sellers, E, Picorelli, R and Salmi, E, 2022. *Comparative Closure: Assessing the biophysical closure challenges of different mining methods,* Western Australia, CRC TiME Report No, Project 3.7.

Svobodova, K, Plieninger, T and Sklenicka, P, 2023. Place re-making and sense of place after quarrying and social-ecological restoration, *Sustainable Development*.

Tavlaki, E, 2015. *RRI Application and Governance.*

van der Ent, A, Baker, A J M, Reeves, R D, Chaney, R L, Anderson, C W N, Meech, J A, Erskine, P D, Simonnot, M-O, Vaughan, J, Morel, J L, Echevarria, G, Fogliani, B, Rongliang, Q and Mulligan, D R, 2015. Agromining: Farming for Metals in the Future?, *Environmental Science and Technology*, 49(8):4773–4780.

von Schomberg, R, 2014. The Quest for the 'Right' Impacts of Science and Technology: A Framework for Responsible Research and Innovation, in *Responsible Innovation 1: Innovative Solutions for Global Issues* (eds: J van den Hoven, N Doorn, T Swierstra, B-J Koops and H Romijn), pp 33–50 (Springer Netherlands: Dordrecht).

Wilson, C E, Morrison, T H, Everingham, J-A and McCarthy, J, 2017. Steering social outcomes in America's energy heartland: State and private meta-governance in Marcellus Shale, Pennsylvania, *American Review of Public Administration*, 47(8):929–944.

Cobalt, copper tailings and the circular economy – a case study from the Capricorn Copper Mine, north-west Queensland minerals province

L Nicholls[1] and A Parbhakar-Fox[2]

1. PhD candidate, Sustainable Minerals Institute, University of Queensland, St Lucia Qld 4072. Email: l.nicholls@uq.edu.au
2. Associate Professor, Sustainable Minerals Institute, University of Queensland, St Lucia Qld 4072. Email: a.parbhakarfox@uq.edu.au

INTRODUCTION

Reprocessing Queensland's historical copper tailings for critical minerals is an opportunity to circularise the state's mining industry by converting waste liabilities into assets. Mine waste valorisation enables economic rehabilitation of legacy mining sites, improves environmental outcomes, and can establish a secondary resources sector that produces a sustainable supply of critical minerals to supplement growing global demands (Australian Government, 2023).

The North-West Queensland Minerals Province (NWQMP) hosts numerous copper mines with known cobalt mineralisation (Degeling, 2020). Cobalt is a critical metal essential for green energy and battery technologies, however its global supply is at risk due to geopolitical and ethical concerns (Rachidi et al, 2021). Historically mining in the NWQMP has focused on copper, leaving cobalt to theoretically accumulate in the mine waste. First-pass geometallurgical characterisation of waste repositories by the Sustainable Minerals Institute identified Capricorn Copper as a key site for further investigation into cobalt deportment and recovery potential.

CASE STUDY SITE

Capricorn Copper is an epigenetic, structurally-controlled copper mine with associated silver and cobalt mineralisation in the NWQMP (Figure 1). Discovered in 1882, the site has both open pit and underground operations on five deposits: Mammoth, Esperanza, Esperanza South, Pluto and Greenstone. The Esperanza tailings dam was operational between 1998 and 2015. Its tailings were primarily sourced from Esperanza (60 per cent), a pyrite-rich orebody hosted in carbonaceous siltstone, and Mammoth (40 per cent) a predominantly supergene copper orebody hosted in quartzite. Ore processing initially involved autoclave leaching with solvent-extraction and electrowinning copper, and later converted to a flotation plant. As with most historical tailings dams, no information is available on spigot location, timing or tailings distribution. Despite this, the Esperanza tailings dam provides a representative case study for assessing the valorisation potential of a historical tailings facility.

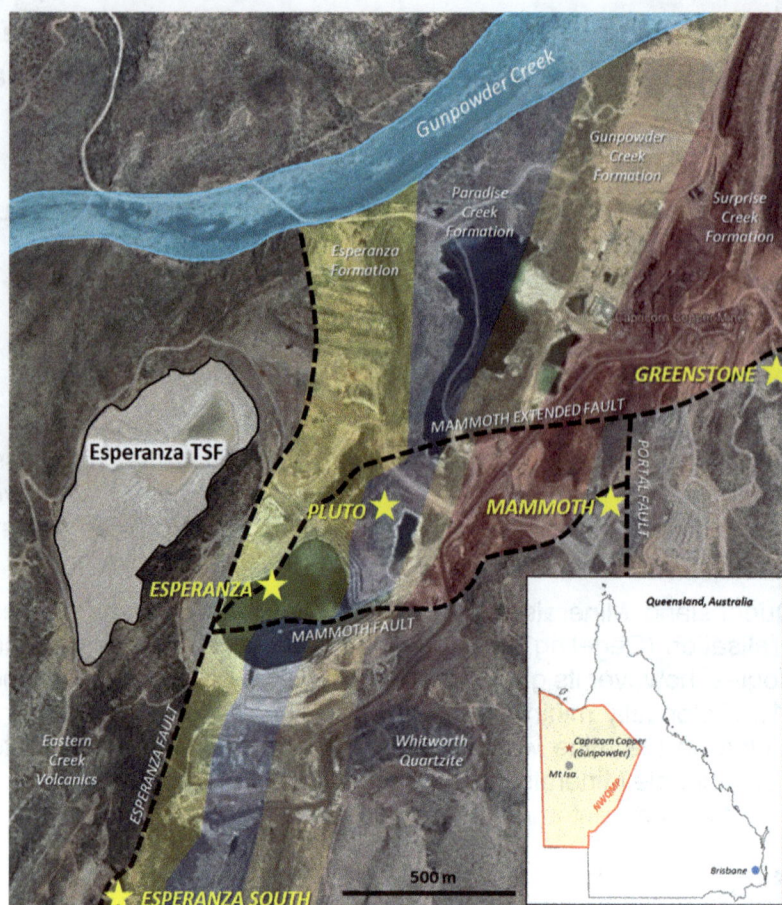

FIG 1 – Capricorn Copper aerial photograph with major geological features. Esperanza Tailings Facility (TSF) highlighted. Inset: Location of mine within the NWQMP.

METHODS

The Esperanza tailings dam was sampled via hand auger on a grid approximately 100 m × 100 m, drilling 26 holes for 232.2 m, down to 10 m depths. Samples were logged by facies, including observations of depth, colour, grain size, saturation, mineralogy and interesting features. 108 representative samples were assayed by 4-acid digest, ICP-MS method at ALS, Brisbane. Of these, 40 samples were sent for mineralogy by X-Ray Diffraction (XRD), and ten subset samples underwent microanalysis by Mineral Liberation Analyser (MLA), Energy Dispersive Spectroscopy (SEM-EDS), microprobe (EPMA) and Laser Ablation ICP-MS (LA ICP-MS). The aim of this research was to investigate the tenor, deportment and distribution of cobalt minerals in the historical tailings and implications these observations have for cobalt recovery.

RESULTS AND DISCUSSION

The historical copper tailings mineralogy consisted mainly of pyrite, quartz, and minor non-sulfide gangue (NSG) minerals such as feldspars, clays, Fe-oxides, sulfates, and carbonates. Minor copper minerals were present, including chalcopyrite, bornite, chalcocite, and covellite, averaging 4.5 wt% of the MLA modal mineralogy. Cobalt occurred in trace (<0.5 wt%) carrollite, alloclasite, and cobaltite, as well as in pyrite. Due to the weathered nature of the tailings, XRD investigations reported up to 50 per cent amorphous material, highlighting the necessity of particle mapping and automated mineralogy using higher resolution SEM equipment for mineral identification.

SEM imagery showed extensive agglomeration of clays and oxides, caused by cyclical wet and dry weathering periods and subsequent oxidation of sulfides (Figure 2). Tailings had an average particle size (p80) of 133 µm, while pyrite grains had a smaller average of 58.42 µm. 45 per cent of pyrite was liberated, with 29 per cent in binary/simple locking and 26 per cent in ternary/complex locking relationships. Cobaltiferous sulfides were extremely fine grained, with an average size of 11.28 µm and 50 per cent liberation. The small grain size and limited liberation pose challenges for conventional mineral processing methods like flotation.

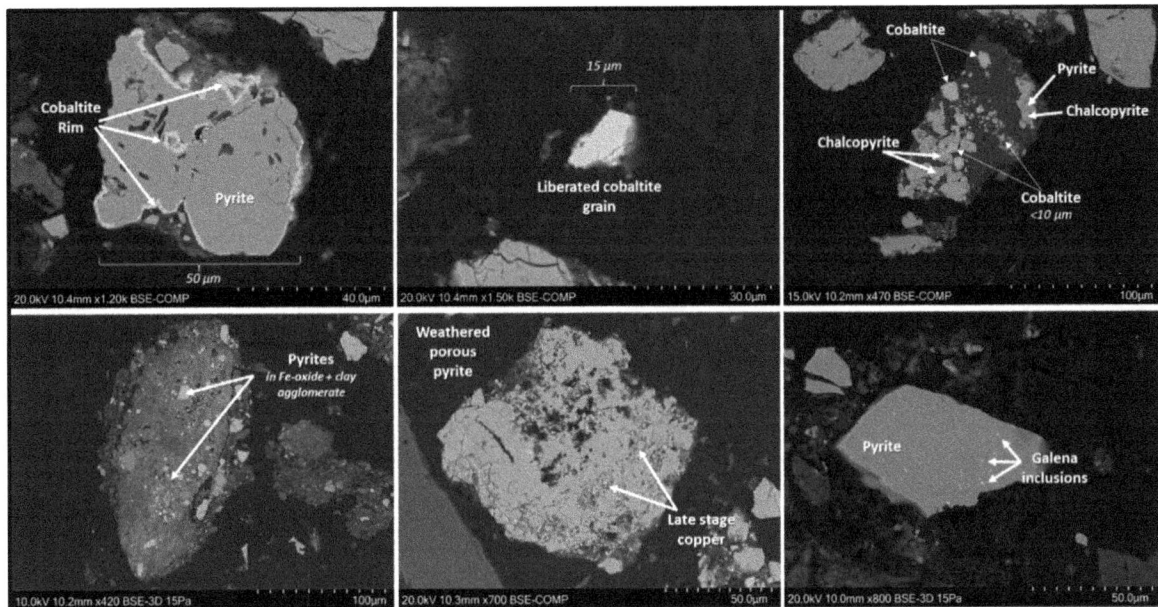

FIG 2 – SEM micrographs of cobaltiferous sulfides and pyrite textures in historical tailings.

Bulk assays revealed cobalt concentrations >1000 ppm associated with both primary sulfide-rich mineralogy and secondary carbonate species with Fe-Mn-Co-Cu substitution. EPMA returned values up to ~3 wt% cobalt in zoned pyrite-cobaltite grains. LA ICP-MS on 1197 pyrite grains returned cobalt values ranging from 1.8 ppm to 8 wt%, distributed across Co-rich zones, in solid solution and as micro-inclusions. As not all pyrites were cobaltiferous, further research is required to discriminate and target specific pyrite phases for cobalt recovery.

CONCLUSIONS

The unique characteristics of historical mine waste and cobalt deportment within the tailings environment have significant implications for mineral reprocessing flow sheet design. When considering flotation to target sulfides, a regrind step or additional agitation may be required to reduce the presence of agglomerates and oxidised surfaces. Fine particle froth flotation may be necessary to recover already fine-grained sulfides, that are further size-reduced from regrinding. Hydrometallurgical processes, such as acid or bioleaching, may be more effective in liberating cobalt from weathered tailings mineralogy with varying mineral species, sulfide states and finer grain sizes. Future metallurgical test work will test these different approaches.

Detailed geometallurgical characterisation on historical tailings in this case study has unlocked key information on cobalt prospectivity in mine waste within the NWQMP and is the initial step to building a circular economy strategy for converting waste liabilities into assets in Queensland.

ACKNOWLEDGEMENTS

Project funding was provided by the Queensland Department of Resources New Economy Minerals Initiative (NEMI). Microanalytical work was conducted at the following institutions: XRD (Central Analytical Research Facility, QUT), MLA (Sustainable Minerals Institute, UQ), EPMA (Centre for Microscopy and Microanalysis, UQ), LA ICP-MS (James Cook University). Many thanks to 29Metals for site access and on-site support.

REFERENCES

Australian Government, 2023. Critical Minerals Strategy 2023-2030 [online], Department of Industry, Science and Resources. Available from: <https://www.industry.gov.au/sites/default/files/2023-06/critical-minerals-strategy-2023-2030.pdf > [Accessed: 10 June 2023].

Degeling, H, 2020. Queensland's new economy minerals initiative [online], *AusIMM Bulletin*, Available from: <https://www.ausimm.com/bulletin/bulletin-articles/queenslands-new-economy-minerals-initiative/> [Accessed: 10 April 2023].

Rachidi, N R, Nwaili, G T, Zhang, S E, Bourdeau, J E and Ghordbani, Y, 2021. Assessing cobalt supply sustainability through production forecasting and implications for green energy policies, *Resources Policy*, 74:102423.

Innovative rehabilitation and closure solutions

High-resolution erosion modelling using the RUSLE – an application to open cut coalmine operations

P Bolz[1], R Chrystal[2] and C M Côte[3]

1. Research Fellow, Sustainable Minerals Institute, Centre for Water in the Minerals Industry, St Lucia Qld 4072. Email: p.asmussen@uq.edu.au
2. Senior Research Fellow, Sustainable Minerals Institute, Centre for Water in the Minerals Industry, St Lucia Qld 4072. Email: r.chrystal@uq.edu.au
3. Associate Professor, Director, Sustainable Minerals Institute, Centre for Water in the Minerals Industry, St Lucia Qld 4072. Email: c.cote@uq.edu.au

INTRODUCTION

The design of final landforms is an integral part of mine closure plans in Queensland and aims to ensure the rehabilitated land is safe, stable and non-polluting. Landform stability in this context is commonly assessed using Landscape Evolutions Models (LEM) which are physical models simulating sediment erosion and deposition around individual landforms over long timespans (up to thousands of years). The most used LEM for assessment of rehabilitated landforms in Queensland is SIBERIA (Willgoose, Bras and Rodriguez-Iturbe, 1991; Tucker and Hancock, 2010).

SIBERIA models are complex, require very detailed input data and simulate erosion and deposition over time frames that extend decades beyond closure. They are not suitable to assess erosion risks during the operational phase of a mine and to support the development of erosion and sediment control plans. Other types of erosion models, such as the Revised Universal Soil Loss Equation (RUSLE, Wischmeier and Smith, 1978) or Water Erosion Prediction Project (WEPP, Nearing *et al*, 1989) calculate soil loss on a monthly or yearly basis and are much better suited to erosion management during the operational phase. They are simpler to implement and can support the design of control measures. For instance, designing a sediment basin requires quantification of sediment loads, which can be produced by these models.

In Queensland, the legislation around sediment and erosion control for open cut mines is currently tightening. Soil erosion remains a major concern due to potential impacts on the receiving environment. It requires management during operation and for closure consideration alike and it is critical that erosion and sediment control plans are developed with the most appropriate tools.

Globally, RUSLE is the most widely used soil loss model (Borrelli *et al*, 2021). This is because it is an empirical model with relatively simple input requirements: biophysical data such as soil properties, vegetation cover, topography and rainfall variability, which can be directly measured or derived from regional or national databases and/or spatial data sets that are available in most parts of the world.

The second most used model is WEPP (Borrelli *et al*, 2021) but compared to RUSLE, there is a large drop in the frequency of its use. This is because WEPP is physically-based and has more complex data requirements. WEPP integrates hydrology, hydraulics, erosion mechanics and land cover to predict soil erosion and sediment transport on hill slope profiles and small watersheds (Stolpe, 2005). It requires some biophysical parameters in formats different from those of RUSLE, as well as detailed hydrologic and catchment data. Nevertheless, it has been used in landform design for mines across Australia to determine final landform stability (eg Howard and Roddy, 2012; Costin, 2020).

In this study, we assess which tool would be the most appropriate to support development of erosion and sediment control plans from operation to closure. We use RUSLE and WEPP to predict soil loss from a mine site, compare the results from both models and outline the advantages of each method.

METHODS

The study area is an operating coalmine with a total catchment area of 335 ha. It was subdivided into 92 sub-catchments using a LiDAR-based Digital Elevation Model (DEM). The HEC-HMS (Hydraulic Engineering Centre – Hydraulic Modeling System (USACE HEC, 2012)) software was used to delineate the sub-catchments of size <40 ha (Figures 1 and 2).

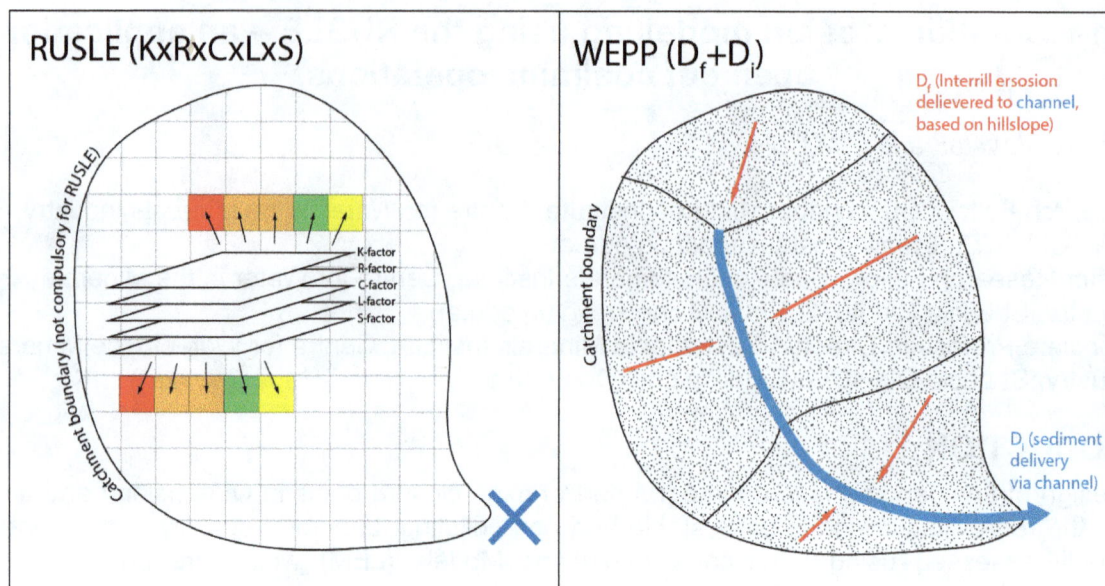

FIG 1 – Conceptualised model structure of the RUSLE (left) and WEPP (right) models. Key differences are pixel-based soil loss for RUSLE versus catchment-based soil loss for WEPP and lack of quantification for sediment delivery for the RUSLE compared to WEPP.

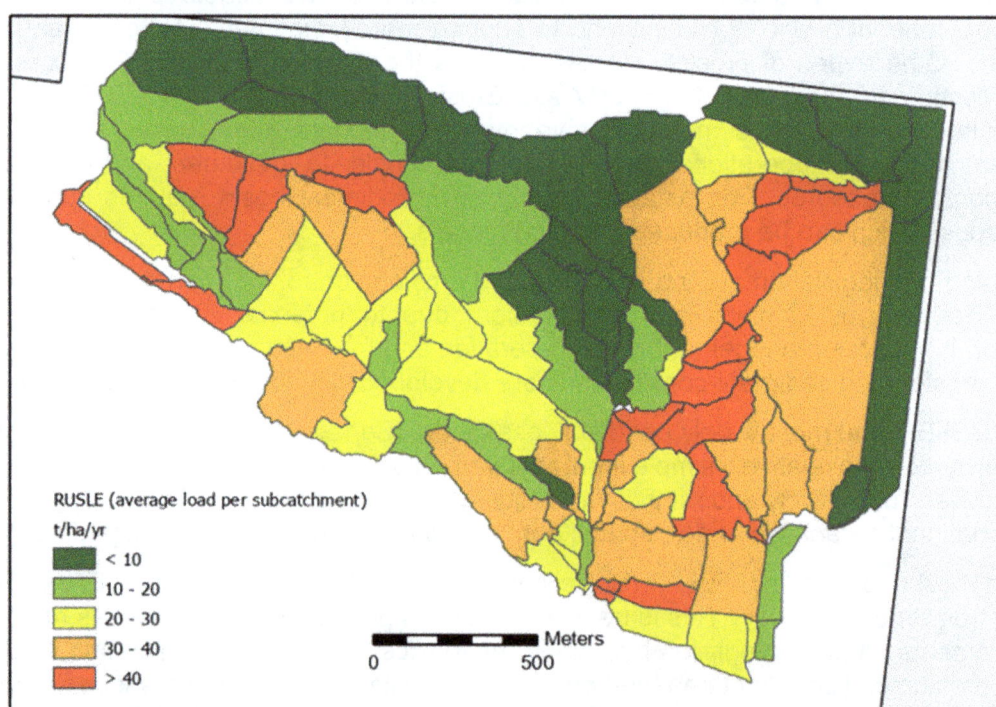

FIG 2 – Average soil loss per sub-catchment calculated by the RUSLE model.

Mine operators regularly acquire detailed topographical data and aerial photography, which can be used to derive vegetation cover. Data sets from mine sites tend to be at a much higher resolution than those obtained from other types of land use. This supports the use of soil erosion models.

Data related to soil, vegetation cover and rainfall were compiled and processed for input into RUSLE and WEPP. Results from both models were compared.

RESULTS

The key distinction between RUSLE and WEPP relates to the structure of each model (Figure 1). RUSLE calculates soil loss for each pixel: in this case, the pixel size was 3 m × 3 m. Soil loss estimates from each pixel are then added to produce the total soil loss from one sub-catchment and this value can be captured on the catchment map (Figure 2). WEPP uses a 'lumped' approach (Wagari and Tamiru, 2021): it calculates the total soil loss for each sub-catchment as one value (Figure 3).

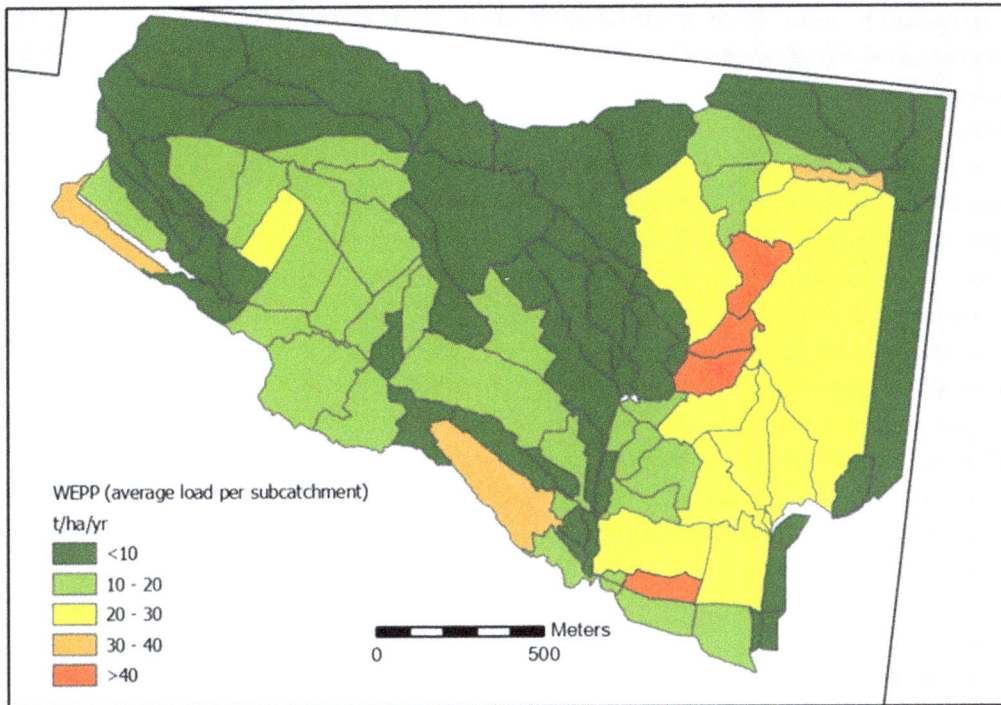

FIG 3 – Average soil loss per sub-catchment calculated by the WEPP model.

Comparison between the two models' results shows good correlation between total soil loss for each sub-catchment (R^2 = 0.88) and average soil loss for each sub-catchment (R^2 = 0.68, Figure 4). Results from RUSLE are generally higher than WEPP, which is in agreement with similar studies comparing sediment yields for these two models (eg Tiware, Risse and Nearing, 2000). This variation can be explained by the difference in models' structure.

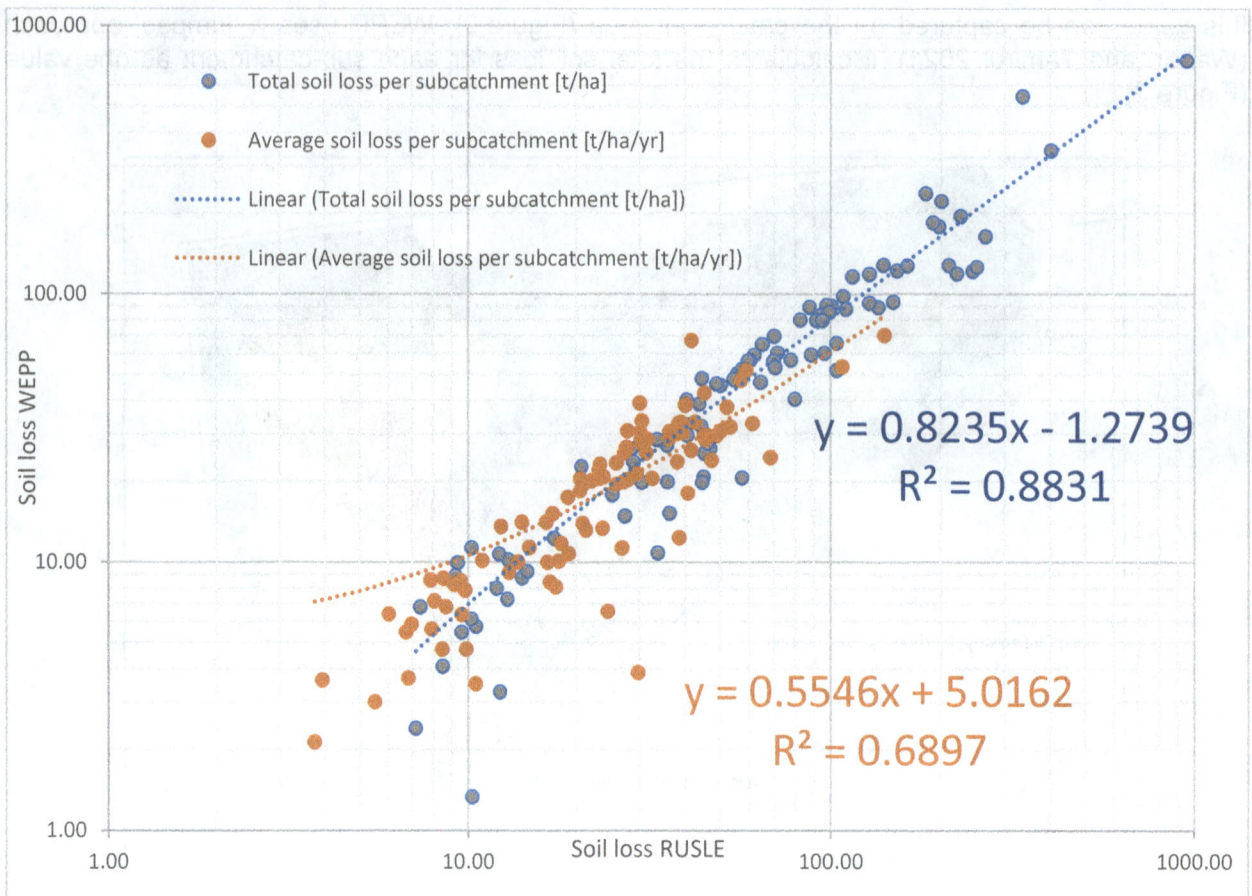

FIG 4 – Comparison of total and average soil loss volumes predicted by WEPP and RUSLE.

Despite this, RUSLE presents the advantage of producing a high-resolution output, which can assist mine sites with targeting areas that most require sediment control. Given the excellent correlation between RUSLE and WEPP results, there does not seem to be any advantage in expanding time and effort towards using WEPP. What RUSLE is missing is the ability to predict sediment transport: this could be addressed through the calculation of sediment delivery ratio (SDR, Walling, 1983), which calibrates sediment delivery at a sub-catchment outlet based on sub-catchment size. In any case, these results lack field validation: this will be studied in a subsequent project.

We have demonstrated that capitalising on the high-resolution data sets acquired by mine sites can support improvement in the design of erosion and sediment control plans, through the calculation of soil loss estimates at a high-resolution spatial scale.

ACKNOWLEDGEMENTS

We would like to acknowledge Yancoal Australia for their past and ongoing support in this project, this work was funded by ACARP Project C29246.

REFERENCES

Borrelli, P, Alewell, C, Alvarez, P, Anache, J A A, Baartman, J, Ballabio, C, Bezak, N, Biddoccu, M, Cerdà, A, Chalise, D, Chen, S, Chen, W, Girolamo, A M D, Gessesse, G D, Deumlich, D, Diodato, N, Efthimiou, N, Erpul, G, Fiener, P, Freppaz, M, Gentile, F, Gericke, A, Haregeweyn, N, Hu, B, Jeanneau, A, Kaffas, K, Kiani-Harchegani, M, Villuendas, I L, Li, C, Lombardo, L, López-Vicente, M, Lucas-Borja, M E, Märker, M, Matthews, F, Miao, C, Mikoš, M, Modugno, S, Möller, M, Naipal, V, Nearing, M, Owusu, S, Panday, D, Patault, E, Patriche, C V, Poggio, L, Portes, R, Quijano, L, Rahdari, M R, Renima, M, Ricci, G F, Rodrigo-Comino, J, Saia, S, Samani, A N, Schillaci, C, Syrris, V, Kim, H S, Spinola, D N, Oliveira, P T, Teng, H, Thapa, R, Vantas, K, Vieira, D, Yang, J E, Yin, S, Zema, D A, Zhao, G and Panagos, P, 2021. Soil erosion modelling: A global review and statistical analysis, *Sci Total Environ*, 780:146494. https://doi.org/10.1016/j.scitotenv.2021.146494

Costin, A, 2020. Landform Evolution Modelling, Isaac River Coking Coal Project, in support of the Progressive Rehabilitation and Closure Plan, Bowen Coking Coal Limited, SGM Environmental Pty Ltd.

Howard, E J and Roddy, B P, 2012, September. Evaluation of the water erosion prediction project model–validation data from sites in Western Australia, in *Mine Closure 2012, Proceedings of the Seventh International Conference on Mine Closure*, pp 81–92 (Australian Centre for Geomechanics: Perth).

Nearing, M A, Foster, G R, Lane, L J and Finkner, S C, 1989. A process-based soil erosion model for USDA-Water Erosion Prediction Project technology, *Transactions of the ASAE*, 32(5):1587–1593.

Stolpe, N B, 2005. A comparison of the RUSLE, EPIC and WEPP erosion models as calibrated to climate and soil of south-central Chile, *Acta Agriculturae Scandinavica, Section B-Soil and Plant Science*, 55(1):2–8.

Tiware, A K, Risse, L M and Nearing, M A, 2000. Evaluation of WEPP and its comparison with USLE and RUSLE, *Transactions of the ASAE, American Society of Agricultural Engineers.*

Tucker, G E and Hancock, G R, 2010. Modelling landscape evolution, *Earth Surface Processes and Landforms*, 35(1):28–50.

US Army Corps of Engineers, Hydrologic Engineering Center (USACE HEC), 2012. Hydrologic Modeling System HEC-HMS, User's Manual, Version 4.0, CPD-74A. Hydrologic Engineering Center (Davis, CA).

Wagari, M and Tamiru, H, 2021. RUSLE model based annual soil loss quantification for soil erosion protection: A case of Fincha Catchment, Ethiopia, *Air, Soil and Water Research*, 14:11786221211046234.

Walling, D E, 1983. The sediment delivery problem, *Journal of Hydrology*, 65(1–3):209–237.

Willgoose, G, Bras, R L and Rodriguez-Iturbe, I, 1991. A coupled channel network growth and hillslope evolution model, 1. Theory, *Water Resources Research*, 27(7):1671–1684.

Wischmeier, W H and Smith, D D, 1978. Predicting rainfall erosion losses: a guide to conservation planning (No. 537), Department of Agriculture, Science and Education Administration.

Polymer-amended tailings dewatering technology for desired closure outcomes

B Boshrouyeh[1], M Edraki[2], T Baumgartl[3] and A Costine[4]

1. Environmental Engineer, O'Kane Consultants, Brisbane Qld 4064.
   Email: bboshrouyeh@okc-sk.com
2. A/Professor, University of Queensland, St Lucia Qld 4072. Email: m.edraki@cmlr.uq.edu.au
3. Professor, Federation University, Churchill Vic 3842. Email: t.baumgartl@federation.edu.au
4. Group Leader, CSIRO Mineral Resources, Waterford WA 6152. Email: allan.costine@csiro.au

INTRODUCTION

In mineral processing, the management of tailings may become increasingly more challenging with a change of ore type and the quality of water used in the plant. This, in turn, may affect the overall costs of the operation and water usage and, in the long-term, pose challenges for mine closure and rehabilitation.

Depending on the applied treatment technique, the short and long-term properties of tailings are largely determined by the fine particles. Most clay materials and ultrafine particles exist in tailings from oil sands, coalmines and base metals, bringing about challenges in dewatering performance. Moreover, given the growing concern for water scarcity, the minerals industry is increasingly turning towards the use of saline water, which poses yet another challenge for tailings management. The high salt content alters the behaviour of clay particles, making dewatering a more complex process than in freshwater systems (Boshrouyeh Ghandashtani et al, 2022; Costine et al, 2018).

High-density techniques for tailings management have been used for several years in the mineral industry. In this regard, one of the novel and promising dewatering methods is in-pipe or inline flocculation, where the polymer is added at high dosages to thickener underflows, giving additional (likely denser) aggregation, releasing more water on deposition compared to low solids thickening.

Previous research in the field of flocculation has primarily focused on the process of flocculation itself, specifically the settling rate and size of aggregates, rather than the long-term impacts of the produced tailings materials (Liu et al, 2020; Jeldres et al, 2017). This is particularly true for the novel method of this technology, where polymer is added to the thickener underflow. Furthermore, these studies have used small-scale set-ups using buckets/beakers, plungers and mechanical stirrers for the flocculation process. Unfortunately, these methods lack the necessary control over the properties of the treated tailings, which can lead to irreproducible mixing and dispersion of viscous polymer solutions (Fawell, Costine and Grabsch, 2015). Consequently, a system that can work with a wide range of operational conditions for high-solid tailings material is essential for reliable and reproducible results.

The objectives of this study were to determine the effect of different type of polymers in the presence of monovalent and divalent salts on the inline flocculation of artificial tailings. The next step was to evaluate the impact of high-dosage polymer addition on geotechnical properties such as settling rate, consolidated density, rate of consolidation and the hydraulic conductivity of tailings. In addition, this research aimed to fill a knowledge gap and investigate the feasibility of using polymer-treated tailing in removing the copper ions and then quantify the performance of produced aggregates under time-dependent processes.

MATERIALS AND METHODS

A synthetic homogenous slurry was used in all experiments to avoid the natural variability that exists with real tailings samples and to enable the preparation of large batches for testing. A slurry was prepared at 50 wt% solids from a combination of 25 per cent kaolin, 45 per cent silica 200G (93 per cent finer than 53 µm) and 30 per cent silica fine sand (99 per cent finer than 0.3 mm). Figure 1 shows the lab-scale experimental set-up used to simulate the inline flocculation process. The main advantage of the low shear chaotic mixer (or topological mixer) as a laboratory tool is that it gives a uniform shear history across the vessel, in contrast to the distinct zones of shear that will exist in a stirred tank.

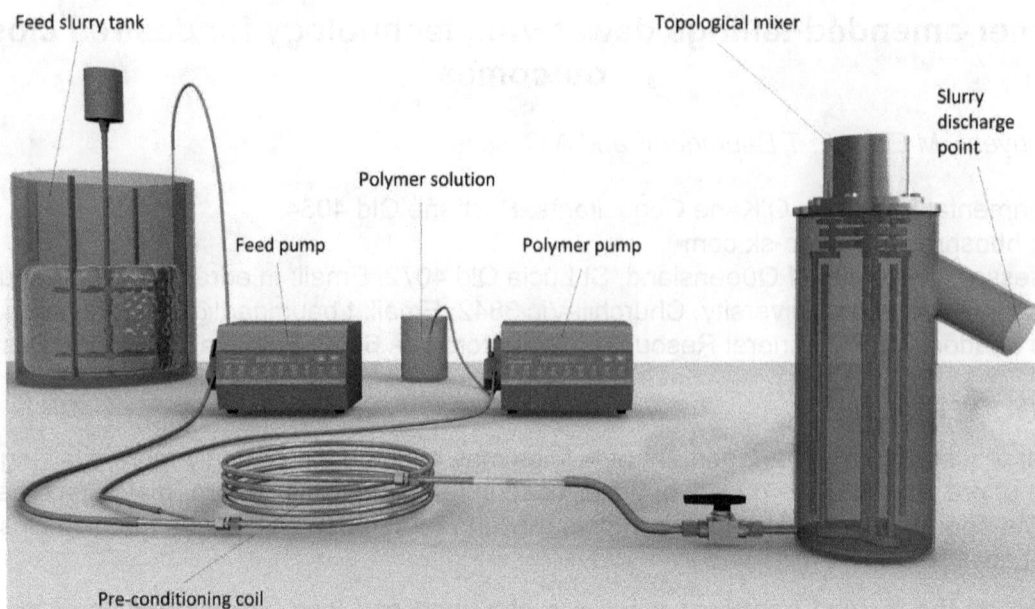

FIG 1 – Tapered shear set-up for polymer addition to high-solids suspensions.

The final products were collected from the discharge point and initially used to assess the effects of elevated concentrations of soluble salts (sodium chloride and calcium chloride) on the inline flocculation process. Additionally, the polymer-treated slurry and its water retention characteristics were evaluated. A slurry consolidometer was then employed to comprehensively monitor the changes in raw and flocculated materials under increasing applied stress. Furthermore, in order to determine the capacity of polymer-treated tailings to remove heavy metal ions, the efficiency of polymer-treated tailings in sorbing/desorbing copper ions from the aquatic environment was examined using the bottle-point method constant concentration technique. Finally, the behaviour of both treated and untreated tailings in settling, evaporation and water drainage over time were assessed in column experiments.

RESULTS AND DISCUSSION

The inline flocculation technique in this study has shown that it is possible to produce the desired highly aggregated network of tailings solids on deposition, even in the presence of elevated concentrations of dissolved salts. This finding has practical implications for mineral processing plants utilising sea water or other brines, and lime addition for pH control in different operations—sodium and calcium-enriched liquors, respectively.

The findings revealed that calcium cations caused a greater reduction in particle separation distances within the tailings slurry. This observation highlights that the interaction between cations in the liquor with exchangeable groups on the mineral surfaces may be better captured in concentrated particle systems. The higher calcium concentration exhibited a more significant adverse impact when combined with low molecular weight copolymers, resulting in aggregates with lower water recovery. Greater compatibility between the high sodium chloride salinity with the high molecular weight copolymer was also exhibited in terms of facilitating the polymer chains' activities to create stronger aggregates.

The results indicated a significant difference in geotechnical and hydrological properties between the raw and polymer-treated slurries, including a different mode of transferring applied stress during consolidation, with 60 per cent less settlement for the polymer-treated sample during the first 48 hrs. Polymer treatment also resulted in a higher void ratio, a lower compressibility index due to relocation of the normal consolidation line, a higher coefficient of consolidation with increased water holding capacity, and 50 per cent more free water drainage compared to the untreated sample. In addition, polymer-treated materials had a higher degree of internal porosity, present as discrete voids rather than an interconnected pore structure.

Polymer treatment resulted in a highly porous structure that exhibited a 25 per cent increased capacity to adsorb and retain copper ions compared to the untreated materials. This behaviour indicates the strong binding between the copper ions and the active site of the treated tailings particles with the increased capability of this material for preserving heavy metal ions within their structure across a wide pH range (2–10) compared to the untreated materials.

Furthermore, the physical changes and longevity of polymer-treated tailings compared to the untreated sample under self-weight consolidation on the atmospheric condition were assessed in this research. A significant difference between the two samples was determined that highlights the impact of the inline polymer treatment of tailings as a sustainable technique. Figure 2 illustrates the development of drying and continuous evaporation/settlement of untreated and polymer-treated tailings over 160 days of monitoring.

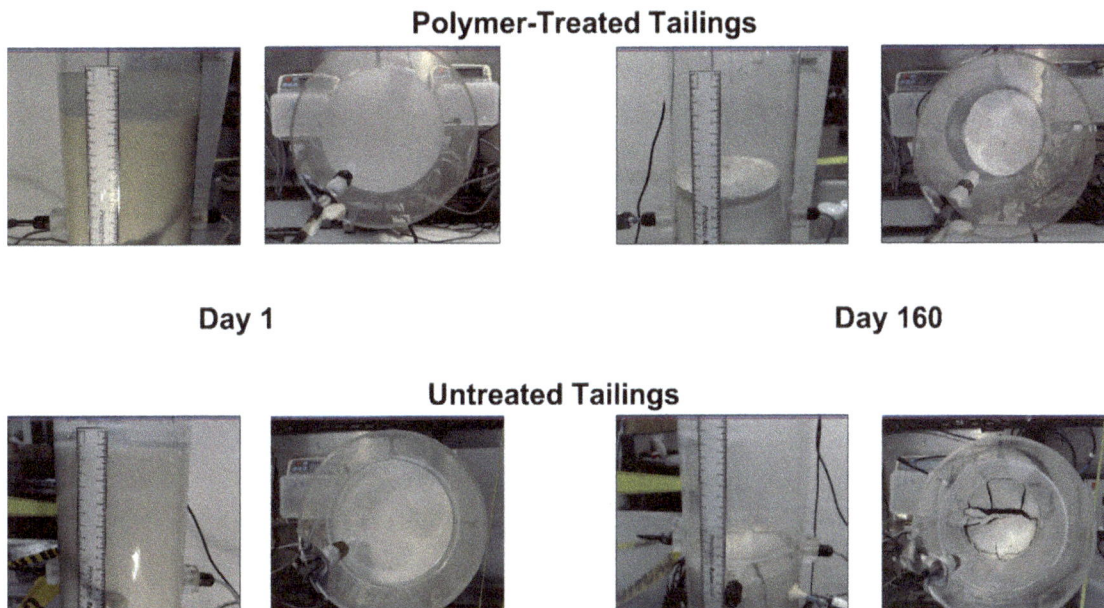

Polymer-Treated Tailings

Day 1

Day 160

Untreated Tailings

FIG 2 – Developing the drying and continuous settlement/evaporation of untreated and polymer-treated tailings samples.

CONCLUSION

The aim of this research was to investigate the physical and chemical stability of tailings aggregates produced through the inline flocculation process using high polymer dosing at high solids content under certain flocculation conditions. In summary, this study provided an understanding of the properties of created aggregates through the inline flocculation technique and their long-term stability under certain conditions. The findings fill several significant knowledge gaps in the current literature and uncover the critical features of inline polymer addition treatment of tailings in favour of potential rehabilitation purposes.

REFERENCES

Boshrouyeh Ghandashtani, M, Costine, A, Edraki, M and Baumgartl, T, 2022. The impacts of high salinity and polymer properties on dewatering and structural characteristics of flocculated high solids tailings, in Journal of Cleaner Production, 342:130726.

Costine, A, Benn, F, Fawell, P, Edraki, M, Baumgartl, T and Bellwood, J, 2018. Understanding factors affecting the stability of inline polymer-amended tailings, in *Paste 2018: Proceedings of the 21st International Seminar on Paste and Thickened Tailings* (eds: R J Jewell and A B Fourie), pp 103–116 (Australian Centre for Geomechanics: Perth). https://doi.org/10.36487/ACG_rep/1805_08_Costine

Liu, D, Edraki, M, Fawell, P and Berry, L, 2020. Improved water recovery: A review of clay-rich tailings and saline water interactions, *Powder Technology*, 364:604–621.

Jeldres, R I, Piceros, E C, Leiva, W H, Toledo, P G and Herrera, N, 2017. Viscoelasticity and yielding properties of flocculated kaolinite sediments in saline water, *Colloids Surfaces A Physicochem Eng Asp*, 529:1009–1015.

Fawell, P D, Costine, A D and Grabsch, A F, 2015. Why small-scale testing of reagents goes wrong, in *Paste 2015: Proceedings of the 18th International Seminar on Paste and Thickened Tailings* (eds: R J Jewell and A B Fourie), pp 153–165 (Australian Centre for Geomechanics: Perth). https://doi.org/10.36487/ACG_rep/1504_10_Fawell

Eight years of ACARP research on rehabilitation of challenging mine sites – practical application in the era PCRPs

G Dale[1], L McCallum[2], A Costin[3], B Silverwood[4] and T Roscoe[5]

1. Managing Director and Chief Technical Officer, Verterra Ecological Engineering, Brisbane Qld 4000. Email: glenn.dale@verterra.com.au
2. Senior Environmental Scientist, Verterra Ecological Engineering, Brisbane Qld 4000. Email: laura.mccallum@verterra.com.au
3. Environmental Engineer, Verterra Ecological Engineering, Brisbane Qld 4000. Email: adam.costin@verterra.com.au
4. Senior GIS and Systems Engineer, Verterra Ecological Engineering, Brisbane Qld 4000. Email: ben.silverwood@verterra.com.au
5. Mining and Energy Sector Lead, Verterra Ecological Engineering, Brisbane Qld 4000. Email: toby.roscoe@verterra.com.au

INTRODUCTION

The introduction of Progressive Closure and Rehabilitation Plans (PCRPs) in Queensland from 2019 places greater importance on planning and implementation of mine rehabilitation activities.

The progress and outcomes of progressive rehabilitation activities is monitored and reported upon to demonstrate success in achieving progress towards the approved post-mining landform, and to inform corrective action where required.

The requirements of PCRPs place additional emphasis on mine sites to quantify the risks of successfully achieving planned outcomes. This is particularly challenging given mined-land rehabilitation is a complex process involving decisions on multiple inter-acting management interventions, climatic conditions and spatially variable site characteristics.

OBJECTIVE

Commencing in 2014, Verterra delivered ACARP project C24033 on best management practices and a decision support tool for rehabilitation of dispersive mine-spoil. ACARP Project C28044 extended this work to mine rehabilitation in general. The objective of this work was to develop a risk-based decision support framework to inform practical, cost-effective management of dispersive mine spoil, together with a set of Best Management Practices (BMPs) and costed, risk-based decision support tools.

THE IMPORTANCE OF SITE CHARACTERISATION

Variability of spoil and topsoil translates to significant variation in site limitations and risk. The importance of reliable site characterisation data to allow site-specific design of amelioration treatments cannot be overstated.

At a minimum, spoil samples should be collected *in situ* following reshaping to enable delineation of spatial variation in spoil properties. Spoil samples should be analysed for parameters that will affect rooting depth and erosivity potential. This will typically include pH, EC, partial size (clay 0–2 µm; fine silt 2–20 µm; coarse silt 20–200 µm and sand 200–2000 µm); and exchangeable cations (Na, Mg, K, Ca). These analytes will inform treatments to address future 'sub-soil' limitations, including pH via either lime or elemental sulfur application to ensure spoil conditions are not hostile to root penetration, and nutrients are available for plant uptake), and dispersity via gypsum application. Application and incorporation of spoil ameliorants should be undertaken prior to topsoiling, ensuring that spoils are treated to a depth of at least 30 cm.

Ideally, topsoil should also be sampled on a grid basis after spreading, but sampling of stockpiles is typically more practical. Topsoil should be analysed for a full physico-chemical suite to inform amelioration of physical, chemical and biological properties, in order to ensure that topsoil can support rapid vegetation establishment and sustain resilient growth.

Grid sampling provides a robust and reliable approach to enable spatial delineation of variation in spoil properties. However, delineation of variation may be assisted by modern digital environmental sensing technologies such as Electromagnetic (EM) and hyperspectral surveying. Typically, a statistical correlation is established between laboratory results and digital survey results at each sampling location. The correlation can then be used to interpolate the results from point samples across the entire area of interest.

INDEPENDENT TREATMENT OF SPOIL AND TOPSOIL

Independent treatment of spoil and topsoil is a novel approach to rehabilitation in mining, however, is logical with demonstrated results in both mining and gully rehabilitation.

Amelioration of spoil limitations prior to application of topsoil allows effective and material-specific treatment to depth of at least 30 cm. Based on a typical topsoil application of 150–200 m, the full depth of ameliorated growing medium post rehabilitation is 450–500 mm. This has been shown to provide two significant benefits:

1. Increased rooting depth (by addressing hostile spoil properties) and hence the volume of stored soil moisture accessible by vegetation. This in turn improves growth vigour and stress tolerance.

2. Increased depth of material resistant to dispersion and erosion. Well-rehabilitated sites have been shown to become more stable over time as applied calcium replaces sodium on the cation exchange complex, and as vegetation becomes established, binding the soil providing raindrop protection and shear resistance.

Erosion issues pre-amelioration, and revegetation outcomes post site-specific soil and spoil amelioration are illustrated in Figure 1.

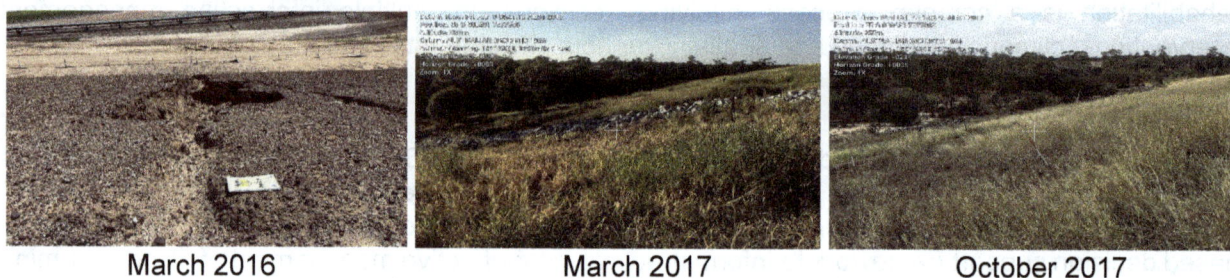

| March 2016 | March 2017 | October 2017 |

FIG 1 – Leading practice soil amelioration and erosion control.

RISK ASSESSMENT

Through reliable site characterisation, rehabilitation treatments can be quantitatively and objectively designed to minimise risk by addressing limitations. Quantitative modelling of rehabilitation risk provides a basis for instilling confidence in PCRPs based on the capacity to implement an objective.

The simplest approach involves spoil and topsoil amelioration to address limitations based on well-established threshold criteria.

The second is to utilise the results of site characterisation to inform spatial application of the revised Universal soil loss equation (RUSLE) (Wischmeier and Smith, 1978). Rehabilitation designs can be determined based on achieving a target maximum level of soil loss.

Third is application of the Bayesian modelling tool developed through eight years of ACARP research (Dale *et al* 2018, 2022). This tool is informed by both the results of site characterisation, but also incorporates management interventions. An optimal rehabilitation design can be achieved by batch analysis of multiple scenarios to identify the combination of treatment and management actions that deliver the least risk outcome for surface and tunnel erosion.

MONITORING FOR CONTROL AND COMPLIANCE

The approach of robust site characterisation carried through to post-rehabilitation monitoring for control and compliance contributes to reduced liability and residual risk, improved rehabilitation performance and increased capacity to meet PCRP requirements. While site characterisation is critical, some areas of broad-scale rehabilitation are always likely to underperform. Early detection of erosion using tools such a dense point-cloud (UAV) LiDAR can effectively identify early-stage erosion and inform early intervention.

Similarly, multispectral surveying enable full site enumeration of vegetation cover and health, informing early targeted intervention, and providing reliable information for compliance reporting.

CONCLUSIONS

A critical learning over the past eight years of ACARP research on rehabilitation of challenging mine sites is the absolute need for reliable, robust and representative site characterisation to inform objective, quantitatively based, repeatable and spatially explicit rehabilitation design. In addition, the information provided by site characterisation may be used to inform all key processes contributing to erosion or poor rehabilitation performance. Finally, the separate treatment of spoil and topsoil, while novel in the industry, has demonstrated proven results by recreating a functional and resilient 'soil' profile.

ACKNOWLEDGEMENTS

Verterra would like to acknowledge the support of ACARP in funding this work and the companies, and their staff that have generously contributed trial and operational sites including Anglo American, Glencore, Peabody Energy, Rio Tinto, New Hope Group and Middlemount Coal.

REFERENCES

Dale, G T, Thomas, E, McCallum, L, Raine, S, Bennett, J Mc and Reardon-Smith, K, 2018. Applying risk-based principles of dispersive mine spoil behaviour to facilitate development of cost-effective best management practices, ACARP Project Report C24033, 194 p.

Dale, G T, McCallum, L, Silverwood, B, Bingham, C, Bennett, J Mc, Ghahramani, A, Ali, A and Robertson, S, 2022. User driven refinement of decision support tools, ACARP Project Report C28044, 294 p.

Wischmeier, W H and Smith, D D, 1978. Predicting rainfall erosion losses – a guide to conservation planning, US Department of Agriculture.

The role of scale in prediction of mine drainage chemistry

M Edraki[1], N McIntyre[2] and K R Jain[3]

1. Associate Professor, Centre for Water in the Minerals Industry, Sustainable Minerals Institute, The University of Queensland, St Lucia Qld 4072. Email: m.edraki@uq.edu.au
2. Professor, Centre for Water in the Minerals Industry, Sustainable Minerals Institute, The University of Queensland, St Lucia Qld 4072. Email: n.mcintyre@uq.edu.au
3. Environmental Engineer, Mine Waste Management, Brisbane Qld 4006. Email: karan.jain@minewaste.com.au

INTRODUCTION

From the very early stages of a mining project, there is a requirement to predict the chemistry of potential drainages from various waste storage facilities. However, using relative masses to scale chemical loads measured by laboratory kinetic tests, such as humidity cells, to a full-scale mine waste facility will lead to concentration predictions that are unrealistically high for many dissolved constituents. Scaling factors developed by various procedures indicate that laboratory rates may range anywhere from two to eight times (Drever and Clow, 1995) to 100 to 1000 times faster than field rates (Malmstrom *et al*, 2000; Smith and Beckie, 2003; Ritchie, 1994). This contribution will use previous works and secondary data to discuss the significance of scale in water quality predictions. It will also present primary experimental geochemical and numerical data from a column leaching study on coal spoils to discuss potential applications of mesoscale leaching tests in bridging the gap between bench-top tests and field scale trials.

METHODOLOGY

Table 1 summarises experimental design and procedures at two scales of column and mesoscale (Jain, Edraki and McIntyre, 2021). Following the derivations in McIntyre, Jain and Edraki (2021), a two-parameter model was developed:

$$dM_m / d_t = -Qc_m + K_1M_{s,im} + K_2M_{s,m} \tag{1}$$

$$dM_{s,im} / dt = -K_1M_{s,im} \tag{2}$$

$$dM_{s,m} / dt = -K_2M_{s,m} \tag{3}$$

where:

t (s)	is time
Q (m^{3}s^{-1})	is the vertical flow rate through the spoil
M_m (g kg^{-1})	is the mass of mobile dissolved salt per unit initial mass of spoil
$M_{s,m}$ (g kg^{-1})	is the mass of solid salt exposed to mobile water per unit initial mass of spoil
$M_{s,im}$ (g kg^{-1})	is the mass of solid salt not exposed to mobile water per unit initial mass of spoil
K_1 (s^{-1}) and K_2 (s^{-1})	are kinetic parameters

Leaching results from two spoils B8 (rock-like) and B9 (soil-like) were selected for scale comparison because they were two distinct types of spoil and had more data than the other spoils across both the scales.

TABLE 1

Experimental design and procedures.

	Code name	Description	Bulk volume (cm³), mass (g) of spoil	Porosity of spoil	Particle size: Sample median and 90 percentile (mm)	Average water volume recovered per cycle (L)	Depth of sample (m)	Average cycle period (days); Number of cycles completed	Wetting-drying regime
Columns	B8 Sat column	Rock-like, saturated, 1-week cycle	2200, 3230	0.45	6.2, 9.2	0.55	0.28	7, 34	Constantly saturated; flowing for three hours every week
	B9 Sat column	Soil-like, saturated, 1-week cycle	2200, 3498	0.48	4.5, 8.8	0.55	0.28	7, 34	
	B8 Biweekly	Rock-like, unsaturated, half-week cycle	1400, 2200	0.41	6.2, 9.2	0.36	0.18	3.5, 34	Constantly unsaturated; water added once every cycle (average addition period 0.3 hours) and gravity drainage collected (average collection period 20 hours)
	B8 Weekly	Rock-like, unsaturated, 1-week cycle	1400, 2150	0.42	6.2, 9.2	0.37	0.18	7, 34	
	B8 Fortnightly	Rock-like, unsaturated, 2-week cycle	1400, 2174	0.41	6.2, 9.2	0.37	0.18	14, 34	
	B9 Fortnightly	Soil-like, unsaturated, 2-week cycle	1400, 1849	0.48	4.5, 8.8	0.35	0.18	14, 12	
Mesocosms	E1 natural IBC	Rock-like, rainfall simulator	1.0×10^6, 1.453×10^6	0.33	46, 192	80	1	266, 8	Unsaturated all the time, 1.5 hours rainfall simulator at start of each cycle
	B8 Sat IBC	Rock-like, saturated	1.0×10^6, 1.552×10^6	0.33	26, 119	318	1	81, 13	Saturated except during gravity drainage at end of each cycle (typical drainage time 0.5 hours)
	B8 unsat IBC	Rock-like, unsaturated-saturated cycles	1.0×10^6, 1.638×10^6	0.33	26, 119	238	1	49, 22	Saturated for three days followed by gravity drainage (typical drainage time 0.5 hours) and unsaturated for rest of cycle
	B9 unsat IBC	Soil-like, unsaturated – saturated cycles	1.0×10^6, 1.616×10^6	0.22	43, 186	126	1	49, 22	

RESULTS AND DISCUSSION

Three phases of salt leaching were observed at both scales:

Phase 1   Leaching of readily available, soluble salts within the first three wetting-drying cycles.

Phase 2   Recession from the peak to quasi-steady conditions within the next ten cycles.

Phase 3   Quasi-steady leaching where slow weathering processes limit salt concentrations.

The salt leaching results from column scale to mesoscale were compared and scale factors derived. There were clear differences in decay rates for column and mesoscale, rock-like and soil-like spoil, saturated and unsaturated leaching, and between different solutes. Table 2, for example, shows decay values of sulfate and calcium per cycle of leaching.

TABLE 2
Decay values of sulfate and calcium (mg/kg/cycle).

			Early stage	Late stage	Early stage	Late stage
			Sulfate		Calcium	
B8 rock-like	Mesoscale	Wet-dry	1.65E-03	2.96E-03	4.96E-05	2.26E-05
	Mesoscale	Saturated	7.81E-03	1.43E-03	1.38E-04	7.95E-05
	Column	Twice/wek	2.52E-02	4.60E-03	1.17E-04	4.01E-05
	Column	Once/week	2.02E-02	5.05E-03	1.25E-04	3.94E-05
	Column	Once/fortnight	1.76E-02	6.75E-03	1.24E-04	4.82E-05
	Column	Saturated	4.77E-02	7.86E-03	3.02E-04	8.25E-05
B9 soil-like	Mesoscale	Wet-dry	2.71E-03	3.59E-03	1.43E-03	1.10E-04
	Mesoscale	Saturated	2.63E-03	4.32E-04	6.87E-04	5.30E-05
	Column	Twice/wek	1.32E-02	4.67E-03	3.15E-03	5.04E-04
	Column	Once/week	2.65E-02	6.09E-03	3.86E-03	5.42E-04
	Column	Once/fortnight	1.93E-02	7.55E-03	3.53E-03	9.95E-05
	Column	Saturated	3.14E-02	5.44E-03	3.58E-03	5.06E-04

The scale factors moving from column to mesocosm under similar moisture conditions and spoil types were 0.13, 0.05, 0.29 (ie the column processes were 7.5, 19, 3.4 times faster) for the slow-process parameter, K_1, and 0.9, 0.23, 0.14 for the fast process parameter, K_2. Similar-in-magnitude scale factors were generally observed when moving from wetter to drier conditions under similar particle sizes and spoil types. Differences in particle size, flow rate, spoil moisture and spoil type influenced scale factor values, although the inter-dependencies between these properties in the experiment create ambiguity. The strong dependence of the K values on flow and moisture conditions encourages:

- The salt decay parameters to be integrated into spoil dump hydrological models.

- The need to characterise spoil dump internal structure and hydrology, especially the degree of heterogeneity in flow and moisture conditions and how this relates to geochemical properties.

ACKNOWLEDGEMENTS

The experimental work was funded by ACARP.

REFERENCES

Drever, J I and Clow, D W, 1995. Weathering rates in catchments, in *Chemical Weathering Rates of Silicate Minerals* (eds: A F White and S Brantley) Reviews in Mineralogy 31, Mineral. Soc. Am, pp 463–481.

Jain, K R, Edraki, M and McIntyre, N, 2021. Controls of wetting and drying cycles on salt leaching from coal mine spoils, *Water, Air and Soil Pollution*, 232(11):472.

Malmstrom, M E, Destouni, G, Banwart, S A and Stromberg, B, 2000. Resolving the scale-dependence of mineral weathering rates, *Environ. Sci. Technol,* 34:1375–1378.

McIntyre, N, Jain, K R and Edraki, M, 2021. Modelling of salt leaching from coal mine spoils at two scales, *Mine Water and the Environment*, 40(4):902–918.

Ritchie, A I M, 1994. Sulfide Oxidation Mechanisms: Controls and Rates of Oxygen Transport, in *Short Course Handbook on Environmental Geochemistry of Sulfide Mine-Waste* (eds: J L Jambor and D W Blowes), Mineralogical Association of Canada: Quebec, 22:201–244.

Smith, L and Beckie, R, 2003. Hydrologic and geochemical transport processes in mine waste rock, *Environmental Aspects of Mine Wastes*, Short Course (eds: J L Jambor, D W Blowes and A I M Ritchie), Mineralogical Association of Canada: Quebec, 31:51–72.

Traceability of native seed for mine rehabilitation

B Fuller[1], C Clarke[2], H Thomas[3] and A Grant[4]

1. CEO, Australian Seeds Authority, Thornbury Vic 3071. Email: bfuller@aseeds.org.au
2. Managing Director, AustraHort, Cleveland Qld 4163. Email: Cameron.clarke@austrahort.com.au
3. CTO, Trust Provenance, Adelaide SA 5073. Email: harley@trustprovenance.com
4. CEO, Trust Provenance, Adelaide SA 5073. Email: andrew@trustprovenance.com

INTRODUCTION

Native seed and plants are often a core input into the rehabilitation process of mine closure. Native plant establishment rate and performance is a function of many input elements including seed quality (purity and germination percentage), processing, treatment, storage, supply chain handling, application and ongoing management.

Traditionally native seed has been supplied with very limited supporting information which in some cases has led to establishment failures, repeated attempts at successful plantings, and biosecurity issues, leading to increased rehabilitation costs for mining companies.

Product procurement scrutiny, and sustainability credential requirements, in all supply chains is increasing the need for transparent, accurate, verified, and certified information. This collectively continues to evolve towards a social license to operate in all industries, including mining.

The Australian Seeds Authority (ASA), in partnership with native seed supplier AustraHort, traceability software company Trust Provenance and supporting partners including Transport for NSW and Queensland Transport and Main Roads (TMR), were recipients of a DAWE National Traceability Grant. The project team has developed a framework and traceability software platform currently being piloted within the Australian native seed industry.

The framework and platform allow for all critical-tracking-events along the journey of a native seed batch to be easily captured utilising a smart phone application (APP) and the global standard for data capture. The platform provides a unique identification for each batch of seed and critical tracking event. The batch ID belongs to that batch and all supply chain stakeholders can contribute to batch traceability by providing more information/data to increase information integrity. This gives the end user, and other stakeholders, increased levels of trust in the product to support value, increase efficiency and reduce costs. For mining companies these efficiencies and cost reductions are in the form of information availability across site/company, potentially improved vegetation establishment, less rework and efficient monitoring. Ultimately this contributes to safe, stable and sustainable rehabilitation.

By capturing data on harvest, logistics, storage (including temperature and humidity), aggregation/blending, treatment, application and ongoing management, the native seed industry and end utilisers of native seed have a greater level of product integrity, brand integrity and information integrity.

How it works

An iPhone and Android APP and web app are used to record critical tracking events as they are undertaken. All data is recorded and accessible via the APP and the website. Each batch of seed harvested is given a unique batch ID. The unique batch ID follows the seed through the supply chain, allowing each stakeholder to add additional critical tracking events to the batch as they occur, such as a cleaning event or a germination and purity certificate (Figure 1). For commonly use critical tracking events used for native seed. The unique batch ID allows for easy inventory management. The app works offline where there is no network and synchronises when back on the network.

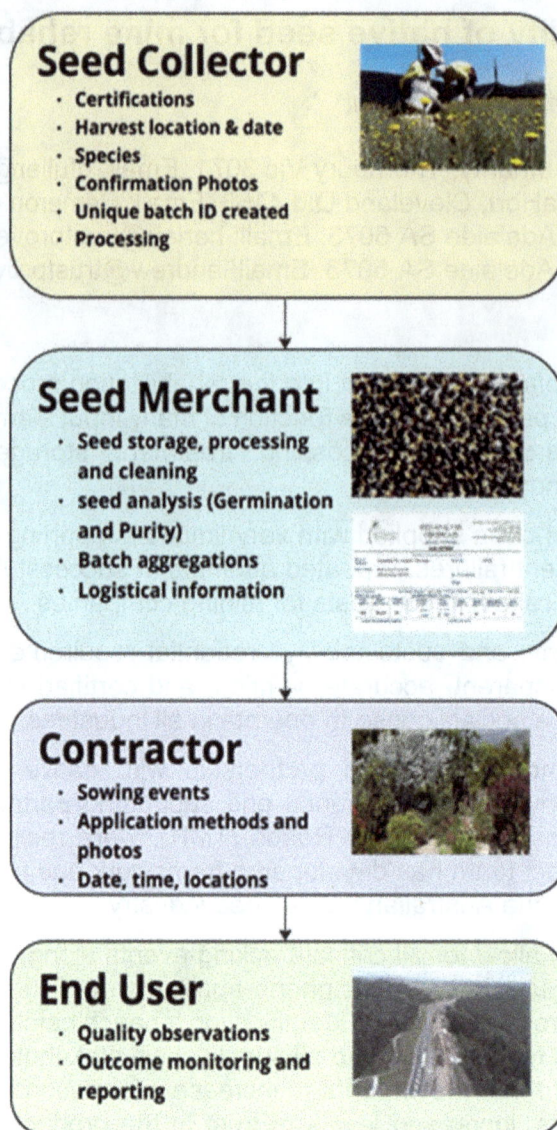

FIG 1 – Critical tracking events captured by stakeholders in a revegetation project.

Supply chain stakeholder uses and benefits

Seed collector and seed merchant

A quick and easy system to manage inventory and traceability. Aligns with growing requirements from customers and supply chain partners for greater product information and integrity. Data capture aligns with industry best practice protocols and guidelines, including *The Australian Guideline for Implementing Food Traceability* 'AGIFT', (a document created via a collaboration across the Australian agricultural and food industry, and Deakin Universities Traceability Lab), and GS1, the global data standard to allow for interoperability between organisations.

All harvest information is captured with permission and/or control of the information collected. System can be used to set reminders for location follow up to time harvesting seasons and as an inventory management tool.

Project delivery manager; eg contractor

All seed coming onto the project site has full traceability. Seed application and post seeding site monitoring of critical tracking events are added to the trace-chain by the contractor/sites.

Provides full detail on all seed across all projects and components of projects. It improves transparency and quality assurance. Aligns with long-term/multiyear projects, to retain data and process integrity. In addition, it mitigates against human resource risk (staff leaving), information loss, contractor changes, harvest partner changes etc.

Example uses for a revegetation project

- The contractor responsible for applying the seed on opening the seed bag is unsure if it is the right seed. By scanning the QR code on the bag, they can see the type of seed, where it was sourced from complete with images confirming species giving the contractor confidence to proceed.

- On inspection of final revegetated area, the company is concerned not all species have established. By scanning the QR code, the company can confirm which species were applied, the seed quality was acceptable and storage conditions along the supply chain would not have affected seed quality. Therefore, the manager can seek out other agronomic issues to improve rehabilitation processes and therefore improve environmental outcomes and accrue the appropriate benefits such as brand integrity and licence to operate.

Project manager (end user)

The Framework provides the ability to outline what seed is required, what areas seed should be procured from and other terms and conditions for harvesting.

For project managers, the traceability framework mitigates risk on quality, process and procurement. As the outcome is process driven it promotes better behaviour. In addition, it can provide inventory management and multiyear project life cycle data capture including monitoring.

The system is complimentary and aligned with other software or data driven applications such as UAV application methods and remote monitoring.

Key benefits for mine site closure and rehabilitation

Traceability of native seed provides mine site closures managers and experts with confidence in seed quality integrity and performance which can lead to transparent information availability across site/company, greater efficiency, opportunities for process improvement (less rework), reduced biosecurity risk, improved rehabilitation success, and enhances speed to rehabilitation certification and potential relinquishment. Ultimately this contributes to safe, stable and sustainable mine rehabilitation.

Contour drains and benches – are they an effective long-term run-off and sediment control method?

G R Hancock[1]

1. Associate Professor, School of Environment and Life Sciences, The University of Newcastle, Callaghan, 2308, Australia. Email: greg.hancock@newcastle.edu.au

INTRODUCTION

A common and cost-effective method for post-mining landscapes is to construct linear hillslopes with a break in the hillslope at appropriate distances to reduce hillslope length and resultant erosion risk (Figure 1). The break in slope occurs at a point that has been found by local experience to be appropriate for the materials and climate. Further enhancement of the linear hillslope may be assisted by surface treatment such as ripping (Saynor, Lowry and Boyden, 2019) (Figure 1). Firstly, ripping introduces surface roughness which breaks up flow paths reducing flow velocity thereby reducing erosion potential. The depression produced by the ripping acts as a trap which captures water, sediment and organic matter (Figure 1).

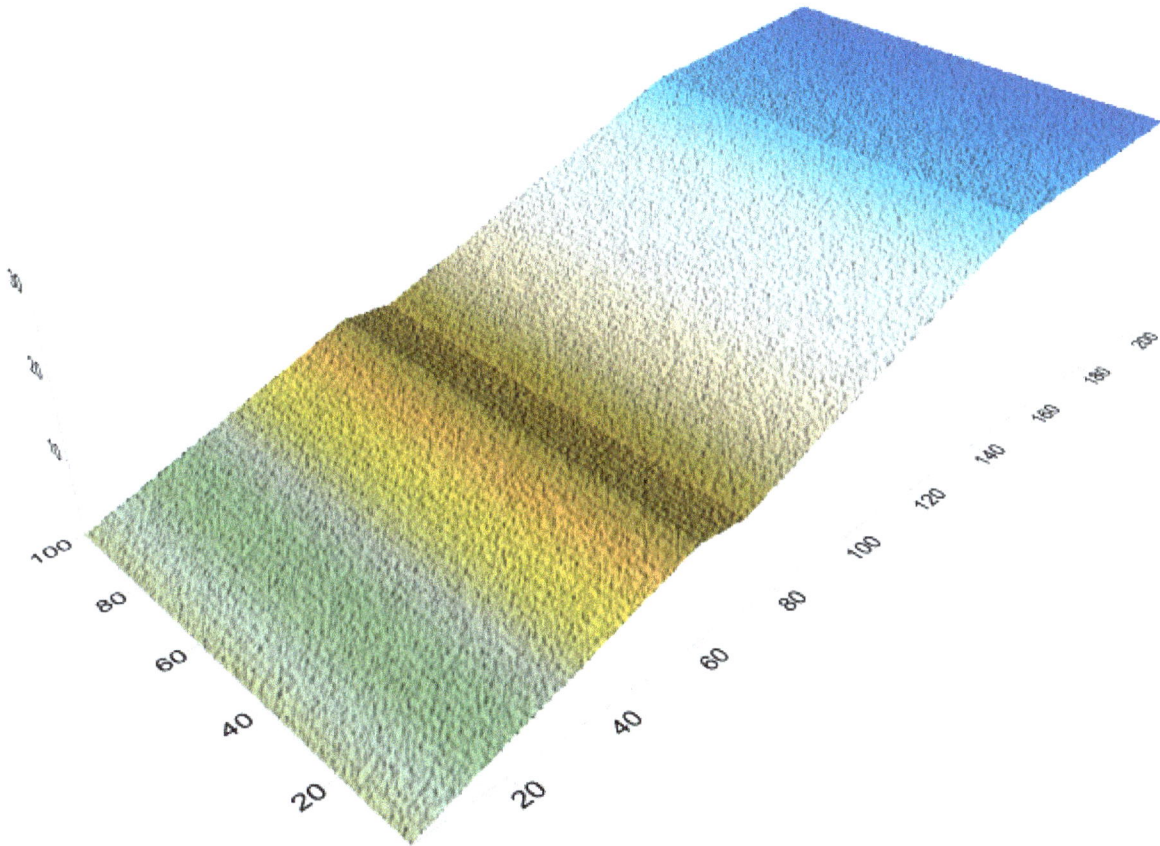

FIG 1 – Hillslope with random roughness and ripping added.

Most areas of the world use linear slopes and benches. However, they can also be prone to failure if inappropriate material are used (Figure 2). There is a need to demonstrate that linear hillslopes will be erosionally stable.

While geomorphic design offers many advantages over linear hillslopes, the designs can be complex and costly to design and construct. Also, there are already many landscapes that have been constructed to a linear design and there is the question as to erosional stability of such landforms at both short and long-term time scales. Should already constructed linear landscapes be reshaped to have an improved hillslope form that reduces erosion and may improve environmental outcomes? Reshaping already constructed landforms will be costly.

FIG 2 – A recently constructed post-mining landform with a large rill and depositional fan at the base of the slope. This occurred immediately post-construction.

The work presented here examines the question as to whether a conventionally constructed hillslope can be erosionally and geomorphically stable at decadal and centennial time scales. Here, a post-mining landscape constructed of linear hillslopes and a bench is examined for its evolution at both decadal and centennial time scales. A LEM (SIBERIA) is used to assess the erosional stability of the landform. A series of different parameters representing a range of climates and surface properties are examined. Different initial surface treatments (ripping and vegetation) are also examined.

MATERIALS AND METHODS

Here a single hillslope is examined which is typical of that used by the mining industry (Figure 1). The hillslope is 200 m long and 100 m wide. The hillslope profile consists of a cap with a slope length of 30 m with a slope of 1 per cent which then increases to 20 per cent for a distance of 80 m (Figure 1). This slope is broken by a bench with a backslope into the hillslope with a slope of 5 per cent.

Two different surface treatments examined are (i) random surface roughness (Figure 1); and (ii) a rip depth of 0.3 m spaced at 1.2 m intervals.

SIBERIA (Willgoose, 2018) is a LEM that has been used extensively for erosion assessment on post-mining landscapes. SIBERIA provides a three dimensional visualisation of erosion and deposition and where it occurs (ie gullies, rills) and rates of erosion and deposition both at the individual grid scale as well as whole landscape (ie t ha^{-1} yr^{-1}).

Before use, parameters for the sediment transport equation and area-discharge relationship are needed. Here a set of existing parameters from a coalmine in the Bowen Basin are used that are considered generally representative. One set of parameters is considered to be 'moderate erodibility' material' and and second set of parameters was consided to be 'low erodibility' material. Both materials are run with and without vegetation.

The SIBERIA model was run for 100 years as this was considered sufficiently long as to provide inference as to longer term landscape trajectory. The 100 year time period is also within a human management time frame.

RESULTS AND DISCUSSION

Moderate erodibility material

At ten years there are small gullies both on the top and bottom lifts with no ripping (Figure 3). There is deposition evident at the base of the top lift on the bench. There is no depositional material at the base of the hillslope as all eroded material is allowed to exit. The gullies are 1.21 m deep and 0.8 to 1 m wide. On the bottom lift there is considerable deposition and reworking of this material.

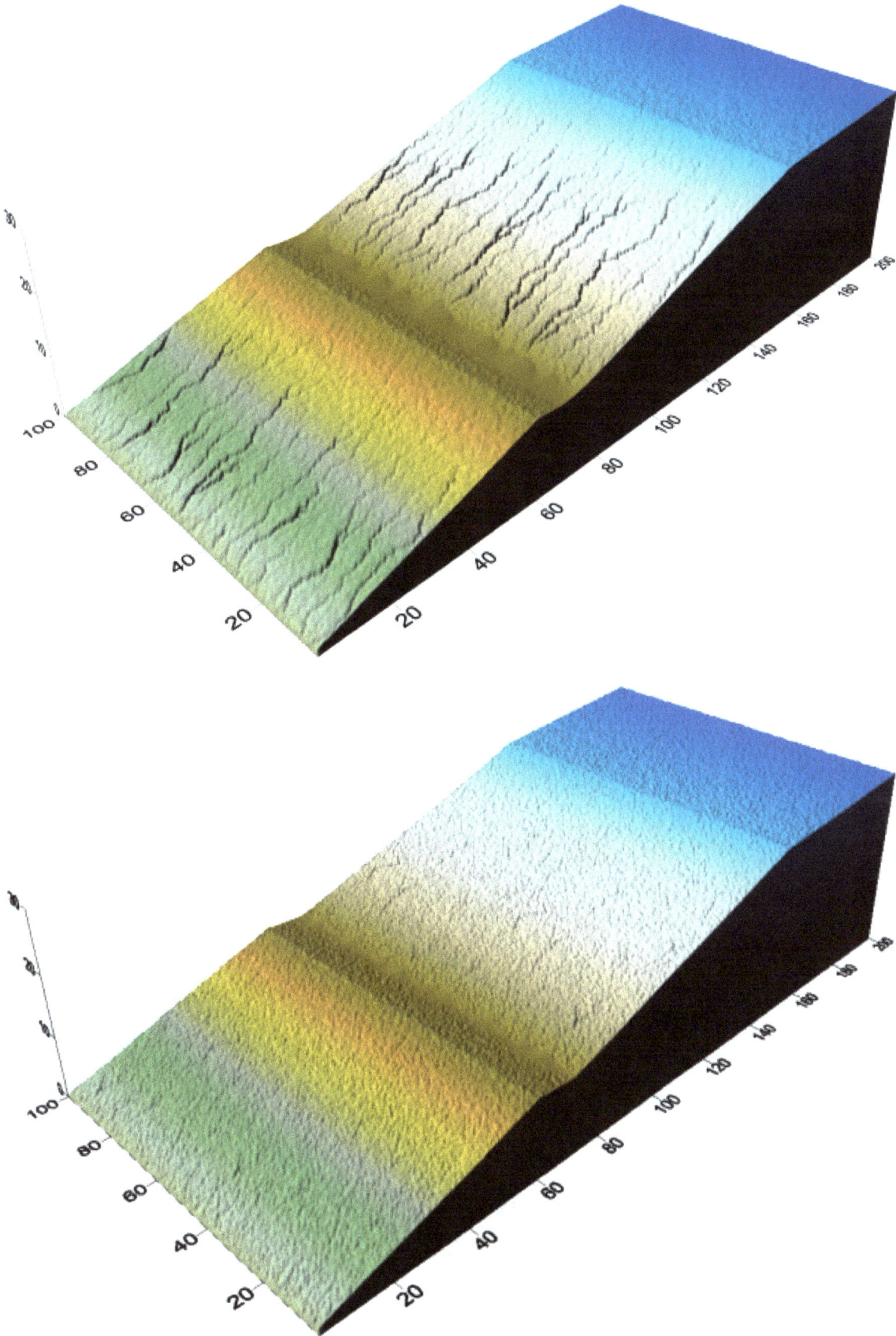

FIG 3 – Post-mine slope at ten years bottom using Moderate Erosion parameters and ripping without vegetation (top) and with vegetation (bottom).

With vegetation, at ten years there is minimal rilling demonstrating the effectiveness of vegetation in managing erosion. Deposition is evident along the bench. However, post-ten years rills and gullies deepen. At 100 years the entire hillslope has gullies with the entire hillslope incised including the cap. Erosion increases once gullying begins (Figure 4). Maximum erosion depth is 0.355 m at ten years and increases to 4.63 m at 100 years.

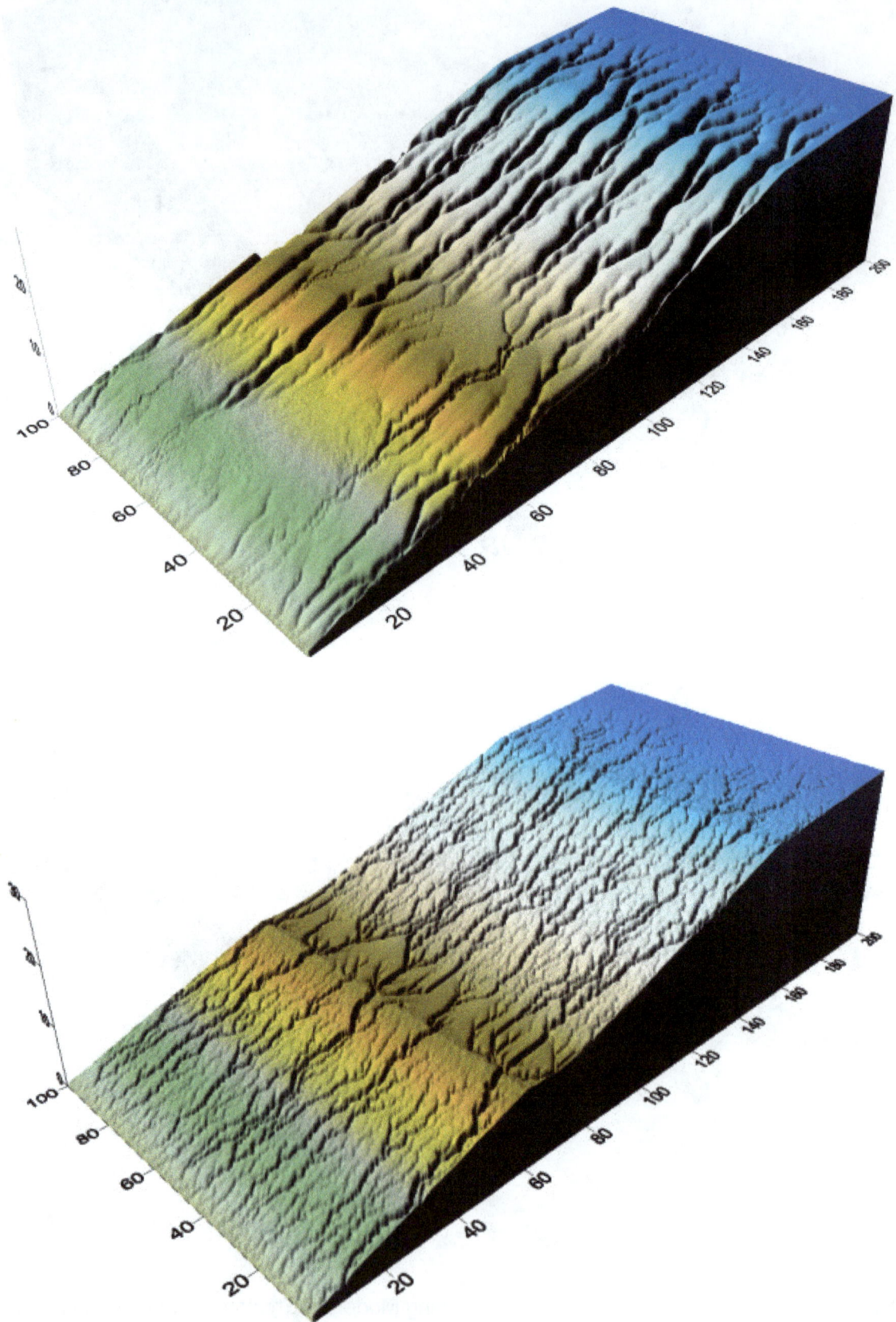

FIG 4 – Post-mine slope at 100 years bottom using Moderate Erosion parameters and ripping without vegetation (top) and with vegetation (bottom).

The results demonstrate that while vegetation may reduce erosion in the short-term (ie ten years), long-term the erodibility of the materials will override the influence of vegetation.

Low erodibility material

Low erodibility material material erosion occurs by sheetwash (Figure 5). Gullies are relatively shalliow and wide. Erosion of the bench has occurred with deposition present but it is still present and effective.

FIG 5 – Post-mine slope at 100 years bottom using Moderate Erosion parameters and ripping with vegetation.

In all cases ripping along the contour reduces erosion over the first ten years. Ripping also prevents gully development. However, the effect of the rips did not last with the erosion rate increasing to that similar to the unripped surface.

CONCLUSIONS

The results demonstrate that a conventionally constructed hillslope can be erosionally stable at centennial time scales. However, this can only be achieved for materials with low erodibility. A landscape should be designed based on the material properties – not make the materials fit the landscape. Over the short-term, forcing the material to fit the landscape may work but over the long-term the materials will override the imposed landscape form. For more erodible materials, vegetation may reduce erosion in the short-term however erosion rates will trend upwards from an initial lower erosion rate.

The major outcome of this work is that material properties matter. A landscape constructed of low erodibility material can be erosionally stable at centennial time scales. That is, conventional landscapes can be erosionally stable with the right material and vegetation. Vegetation may reduce erosion for higher erodibility materials however vegetation may fail, be reduced by drought and fire or poorly managed (ie introduction of grazing animals). Vegetation alone is not a long-term solution for erosional stability.

Ripping offers the potential to reduce erosion over the short-term. However, the effectiveness of ripping reduces over time and the material properties come to dominate over rather than than the surface form. Ripping is a short-term solution to mitigate erosion for materials with low erosional stability but it is not a long-term solution. Ripping will not reduce erosion for high erodibility materials over the long-term. Ripping may lead to a false confidence in the early years post construction with failure occurring in 5–10 years.

ACKNOWLEDGEMENTS

This work was supported by the mining industry through Australian Coal Association Research Program (ACARP) Projects C34025: New Landscape Evolution Model for Assessing Rehabilitation and C27042: Adaption of design tools to better design rehabilitation and capping over highly mobile mine waste and Australian Research Council Discovery Grant DP110101216.

REFERENCES

Saynor, M J, Lowry, J B C and Boyden, J M, 2019. Assessment of rip lines using CAESAR-Lisflood on a trial landform at the Ranger Uranium Mine, *Land Degrad Dev*, 30:504–514. https://doi.org/10.1002/ldr.3242

Willgoose, G R, 2018. *Principles of Soilscape and Landscape Evolution* (Cambridge University Press: Cambridge).

Diavik Diamond Mine north country rock pile – an example of integrated mine planning for successful progressive reclamation

V Knott[1], S Sinclair[2], A Yeomans[3], G MacDonald[4] and G Stephenson[5]

1. Engineering Superintendent, Projects, Diavik Diamond Mine, Yellowknife, Northwest Territories, Canada, X1A 2P8. Email: veronica.knott@riotinto.com
2. Principal Advisor, Closure Planning and Design, Diavik Diamond Mine, Yellowknife, Northwest Territories, Canada, X1A 2P8. Email: sean.sinclair@riotinto.com
3. Senior Design Engineer, Closure, Diavik Diamond Mine, Yellowknife, Northwest Territories, Canada, X1A 2P8. Email: arthur.yeomans@riotinto.com
4. Manager, Closure, Diavik Diamond Mine, Yellowknife, Northwest Territories, Canada, X1A 2P8. Email: gord.macdonald@riotinto.com
5. Manager, Surface Operations and Projects, Diavik Diamond Mine, Yellowknife, Northwest Territories, Canada, X1A 2P8. Email: gord.stephenson@riotinto.com

INTRODUCTION

Rio Tinto has actively committed to progressive reclamation, concurrent with commercial operations, at its Diavik Diamond Mine. The most significant piece of progressive reclamation undertaken at Diavik to date has been the construction of an engineered thermal rock cover for the mine's largest waste rock storage area. This 80-metre-tall and 2000-metre-long area is known as the Waste Rock Storage Area – North Country Rock Pile (WRSA-NCRP) and contains the operationally segregated potentially acid-generating waste rock mined over the last 20 years.

As part of the mine plan, an opportunity was recognised where a limited 1.5 per cent of the WRSA-NCRP surface area was required for storage of the balance of life-of-mine generated waste rock and a decision was made to advance the closure design almost a decade ahead of final closure. With over 12 860 000 tonnes of material placed and 57 443 dozer hours logged since cover construction, progressive reclamation on the WRSA-NCRP is scheduled to be complete in mid-2023, three years before planned cessation of commercial operations in 2026.

The progressive closure of the WRSA-NCRP demonstrates Rio Tinto's commitment to not only closure, but modern best practices with integrated mine closure. Over the past seven years, Diavik has overcome multiple implementation hurdles for this project including the psychology of operational standby, the financial mechanisms required to support closure, working within the constraints of a mine plan and the associated operational organisational structures. This paper outlines the lessons learned to provide an example of successful integrated mine planning for closure.

The North Country Rock Pile project had three major learnings which resulted in an effective organisational structure that allowed for the execution of an integrated planning model for progressive closure.

The initial hurdle the project faced was adapting the psychology of an operational open pit mining team. The team was an experienced mining group who came in with the perspective to target the shortest haul possible resulting in the initial perception that the longer haul of material to the NCRP was an un-necessary task impacting operational performance metrics. After the installation of an improved dispatch tracking system and performance metrics, it was possible to demonstrate that the mine was shovel constrained. This showcased the short haul was generating an increase in queue time not production. With updated performance metrics, the team was aligned to target a decrease standby through haulage to the NCRP.

These performance metrics had to be reinforced by financial mechanisms. The operations team was able to offset costs by charging time to the capital project set-up for the closure of the NCRP. This allowed the operations team to receive credit for time that would have otherwise likely been standby. As well, the progressive closure project was able to receive the benefit of not having to incur costs for the drill, blast and load as that was done as part of the operation. This was a significant benefit to the project and one of the primary drivers for the early approvals of the work as the project costs were limited to the incremental haulage distance and placement of the material.

While the combined operational and progressive closure approach provided financial advantages, it did result in the project being constrained by the operational activities. A traditional capital project is managed through a linear execution plan where the deliverables within a project are within the realm of control of the project execution team. However, the NCRP project had to be continuously adapted to the outcomes of the pit operations. This included geotechnical constraints, ore delivery, and other operational delays. To adapt, the planning module had to be dynamic but not reactive to allow for adjustments on a shift-by-shift basis. Based on this requirement the Diavik Closure Projects structure adopted a mine planning format. The team executed a similar planning timeline and reporting structure as the mine operations team, allowing planning to be integrated into the operational planning structures on a daily, weekly, monthly and quarterly basis. This resulted in the Projects team as a core pillar of the mine planning team.

The process of continuous improvement spanned six years and has resulted in the project being ahead of schedule and under budget for the project's scheduled progressive closure completion date of June 2023. Figure 1 showcases the rock fill placement as of December 31, 2022, covering 96 per cent of the rockpile.

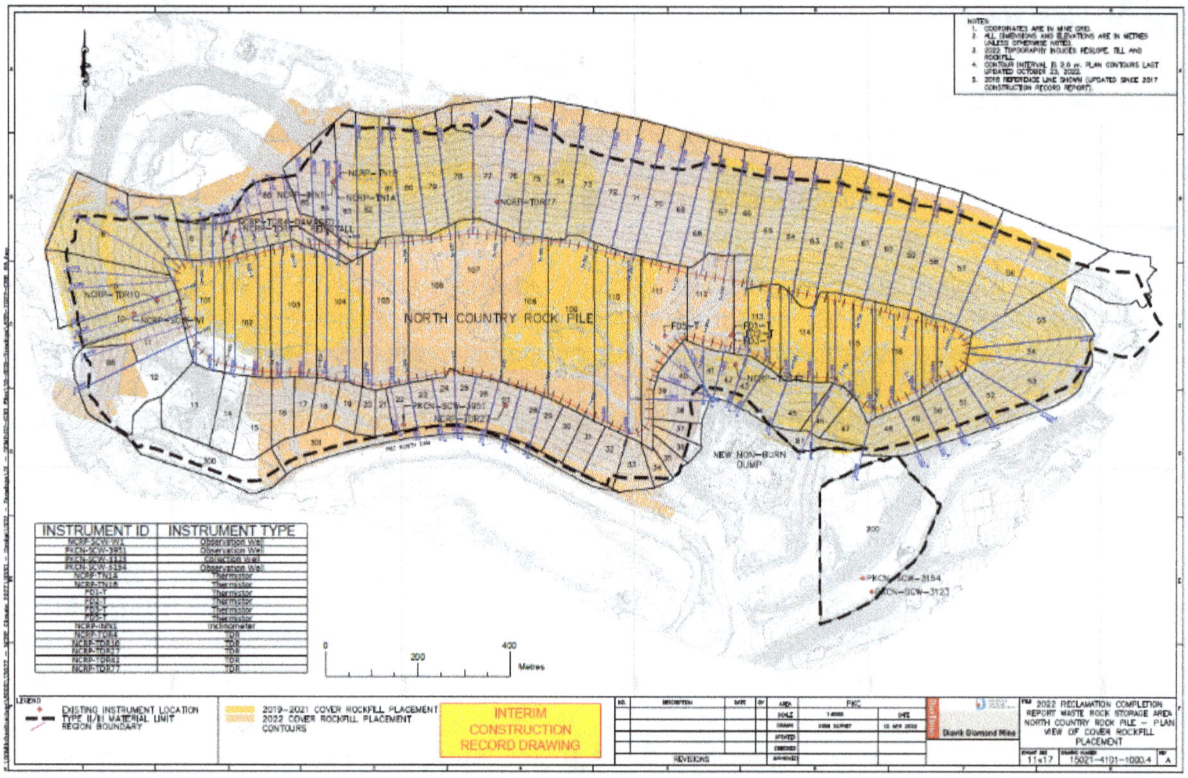

FIG 1 – Rock fill cover placement as of Dec 31, 2022.

The North Country Rock Pile project supports the argument that bulk earthworks progressive closure projects should be integrated into the operational mine plan. If the project is aligned the mine plan it allows for dynamic adjustments which capitalise on evolving opportunities inevitable in an operating mine. These cultural and organisational alignments should be supported by appropriate performance metrics and financial mechanisms to ensure united priorities. The result of this alignment provides long-term financial savings for the business as well as delivery of closure objectives.

ACKNOWLEDGEMENTS

The authors would like to thank Rio Tinto Diavik Diamond Mines who contributed by providing direction, input and support in the development of this work.

Assessing the effectiveness of erosion predictions from a landform evolution model with field observations from a rehabilitated landform

J B C Lowry[1], M J Saynor[2], G R Hancock[3] and T J Coulthard[4]

1. Landform modeller, Supervising Scientist Branch, Department of Climate Change, Energy, the Environment and Water, Darwin NT 0801. Email: john.lowry@dcceew.gov.au
2. Senior Research Scientist, Supervising Scientist Branch, Department of Climate Change, Energy, the Environment and Water, Darwin NT 0801. Email: mike.saynor@dcceew.gov.au
3. Associate Professor, School of Environmental and Life Sciences, The University of Newcastle, Callaghan NSW 2308. Email: greg.hancock@newcastle.edu.au
4. Professor, Energy and Environment Institute, University of Hull, Hull HU6 7RX, United Kingdom. Email: t.coulthard@hull.ac.uk

INTRODUCTION

Landform evolution models (LEMs) can assess the effectiveness of landform designs applied to a rehabilitated landform by predicting where erosion may occur. Here, we compare erosion predictions produced by the CAESAR-Lisflood LEM with field observations from the rehabilitated Pit 1 landform of the Ranger uranium mine, based in the Northern Territory, Australia (Figure 1). The ability to predict long-term erosion at a catchment scale is important, as the final rehabilitated Ranger landform is required to physically isolate buried tailings material for at least 10 000 years.

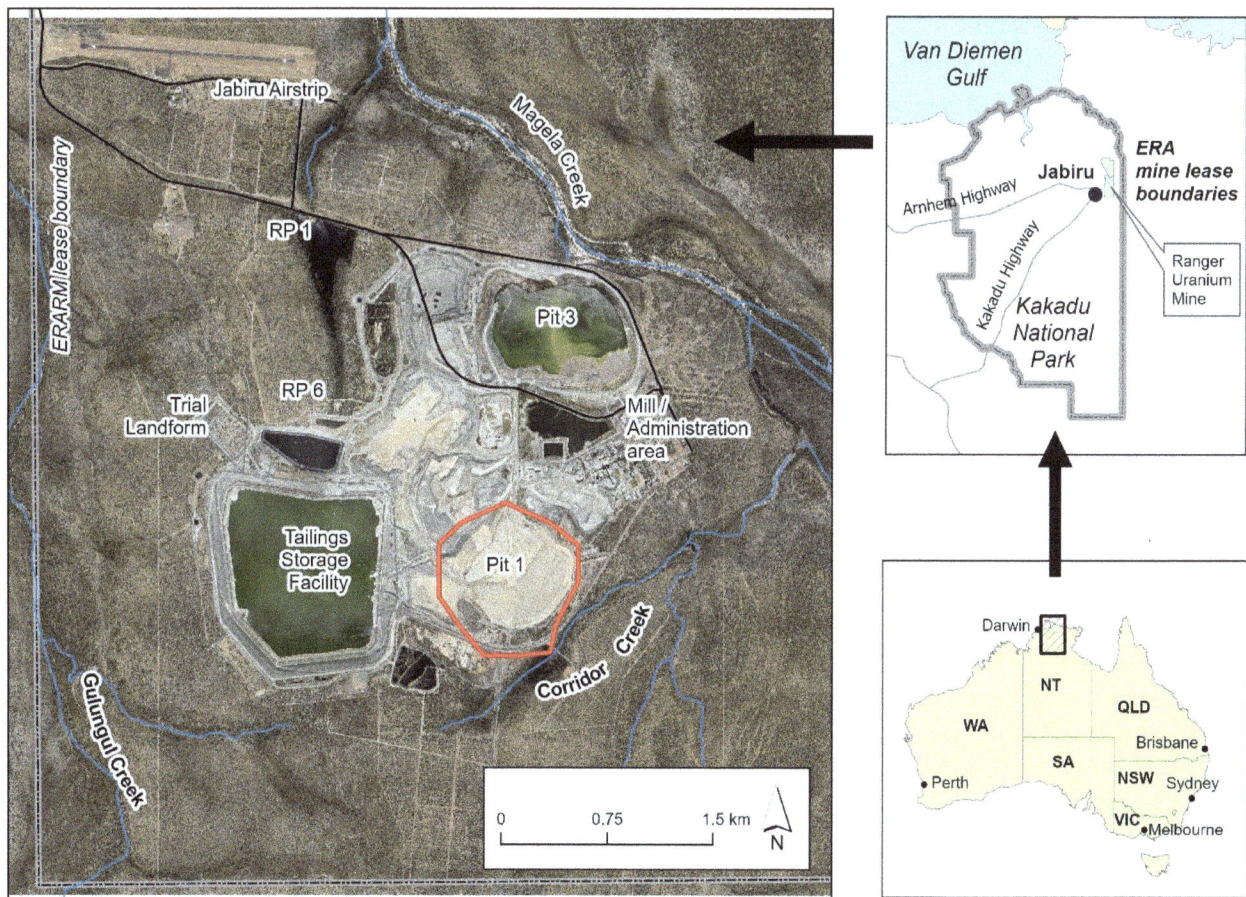

FIG 1 – Aerial mosaic from 2020 of Ranger mine with the rehabilitated Pit 1 landform highlighted in red.

CAESAR-Lisflood has been extensively tested and calibrated using data collected from erosion plots (0.09 ha) on the Ranger trial landform (TLF) for multiple years (Lowry et al, 2020; Saynor, Lowry and Boyden, 2019). However, until recently, it was impossible to compare model predictions with field observations for areas larger than the TLF erosion plots. We can now compare model predictions of gully development with field observations from the newly rehabilitated, 40 ha surface of Pit 1 over two contiguous one-year periods: 2020–2021 and 2021–2022. The aim of this study was to assess whether the CAESAR-Lisflood model can successfully predict gully erosion at a larger spatial scale.

METHODS

This study utilised the CAESAR-Lisflood v1.9j LEM. It utilised three key data inputs:

1. A digital elevation model (DEM).
2. Rainfall data.
3. Surface particle size distribution data.

DEMs of the Pit 1 surface were generated from imagery acquired by remotely-piloted aircraft system (RPAS) surveys flown in October 2020 and in September 2021. All DEMs were resampled to a horizontal resolution of 1 m and were referenced against the Geocentric Datum of Australia 1994 (GDA94) and Australian Height Datum, to ensure consistency between model outputs and the products generated from RPAS surveys.

Model simulations were run as two separate one-year periods, and utilised rainfall data collected from 1 September 2020 to 31 August 2021 and from 1 September 2021 to 31 August 2022 on the nearby Ranger trial landform.

Particle size data collected on the trial landform surface (Hancock *et al,* 2020), were used to represent the surface particle size classes used in model simulations.

For the purposes of this study, CAESAR-Lisflood simulations used the Wilcock and Crowe (2003) sediment transport equation.

Regular inspections of the Pit 1 landform, recording the position of gully features were conducted, commencing from October 2020. Aerial surveys were undertaken at approximately 2-monthly intervals with a camera-equipped DJI Phantom 4 Pro V2.0 RPAS to generate high-resolution (20 cm) orthomosaic images and DEMs of the pit surface.

For the 2020–2021 rainfall year, imagery captured on 26 October 2020 was used as the initial DEM input to CAESAR-Lisflood. Imagery captured on 10 February 2021 was used for comparison purposes of the post-gully landform. For the 2021–2022 rainfall year, imagery captured on 1 September 2021 was used as the initial DEM for input to CAESAR-Lisflood. Imagery captured on 2 March 2022 was used for comparison purposes of the post-gully and post-wet season landform.

ESRI ArcGIS Pro v8.3 software was used to compare differences between the DEMs generated from the RPAS flights and by the CAESAR-Lisflood LEM. The assessment of the DEMs involved comparing the distribution, extent and depth of gullying mapped or predicted for each rainfall year.

RESULTS

Year 1: 2020–2021

Analysis of the aerial imagery captured over Pit 1 on 10 February 2021 identified the presence and location of the two gullies on the edge of the Pit 1 landform (Figure 2). No additional gullies were identified from the imagery.

The CAESAR-Lisflood model simulations for the 2020–2021 rainfall year predicted the occurrence of two gullies on the surface of Pit 1 in the same time period and in the same area on the landform (Figure 3).

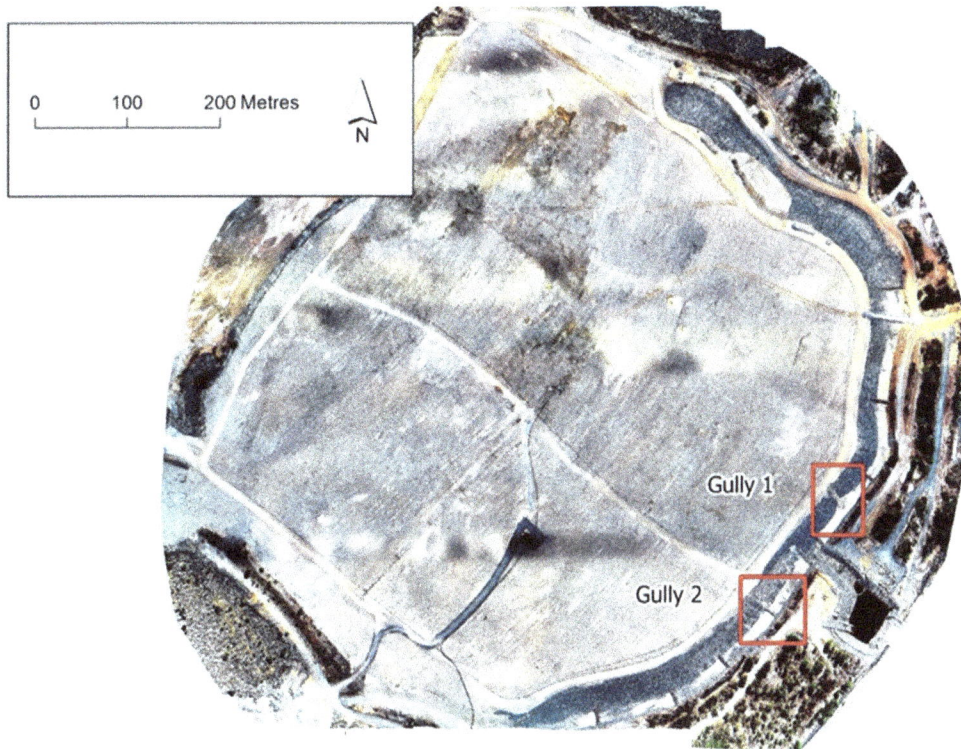

FIG 2 – Location of two gullies (in red boxes) observed from 10 February 2021 flight.

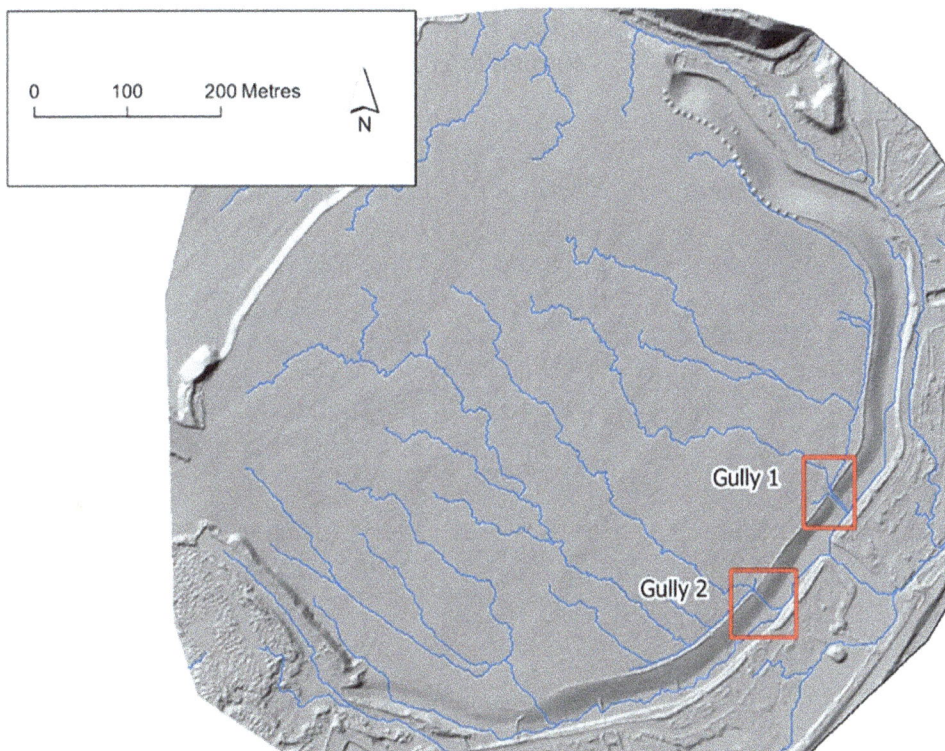

FIG 3 – Model predictions of gullies (red boxes) and drainage network (blue lines).

Year 2: 2020–2021

Further observation of the Pit 1 surface showed that the two gullies that had formed in 2020–2021 and had been repaired, had re-formed by March 2022. In addition, three additional gullies (five in total) were observed to have formed on the eastern edge of the Pit 1 landform. The location of these gullies was identified in aerial imagery captured in March 2022 (Figure 4).

The CAESAR-Lisflood simulations for the 2021–2022 rainfall year matched the field observations of gully formation on the Pit 1 surface. The modelling predicted the formation of the five observed gullies with two in the same location in which gullies had formed in 2021 (Figure 5). An additional gully (Gully 6) was predicted by the model to occur but was not observed to have formed on the Pit 1 landform.

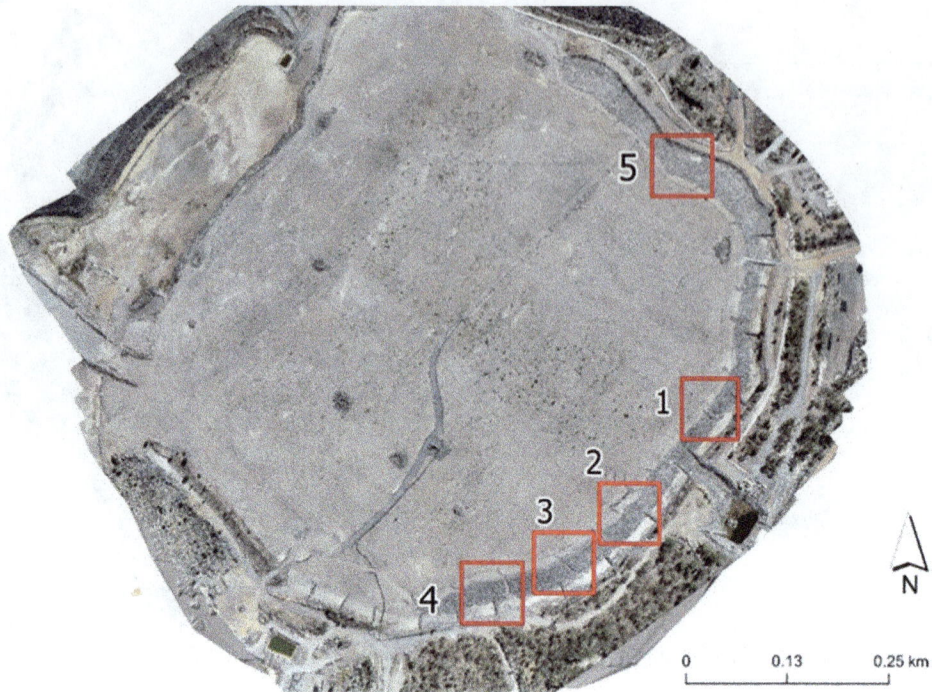

FIG 4 – Location of gullies on the surface of Pit 1 – 2 March 2022. Gullies 1 and 2 also formed in 2021.

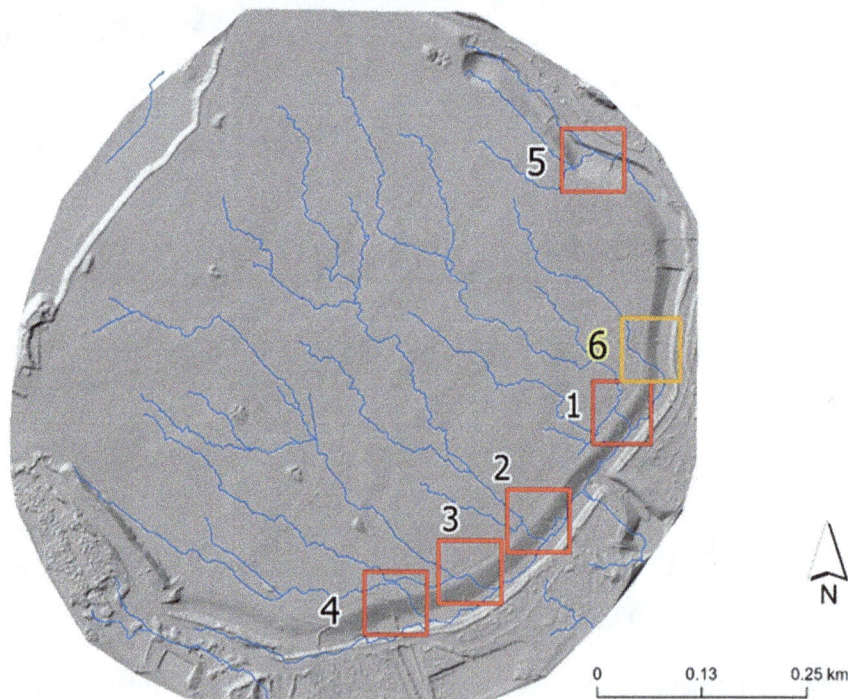

FIG 5 – Predicted gully locations March 2022 (an additional gully location (#6) was predicted by the model).

Overall, the results demonstrated that the model was able to successfully predict the location of gullies in the second wet season (five observed with six predicted). Cross-section profiles across a DEM generated from a RPAS flight flown after the gullies formed were compared with profiles generated from the DEMs generated by CAESAR-Lisflood. These showed that CAESAR-Lisflood predicted gullies to occur in similar locations to those observed from the RPAS flight (Figure 6).

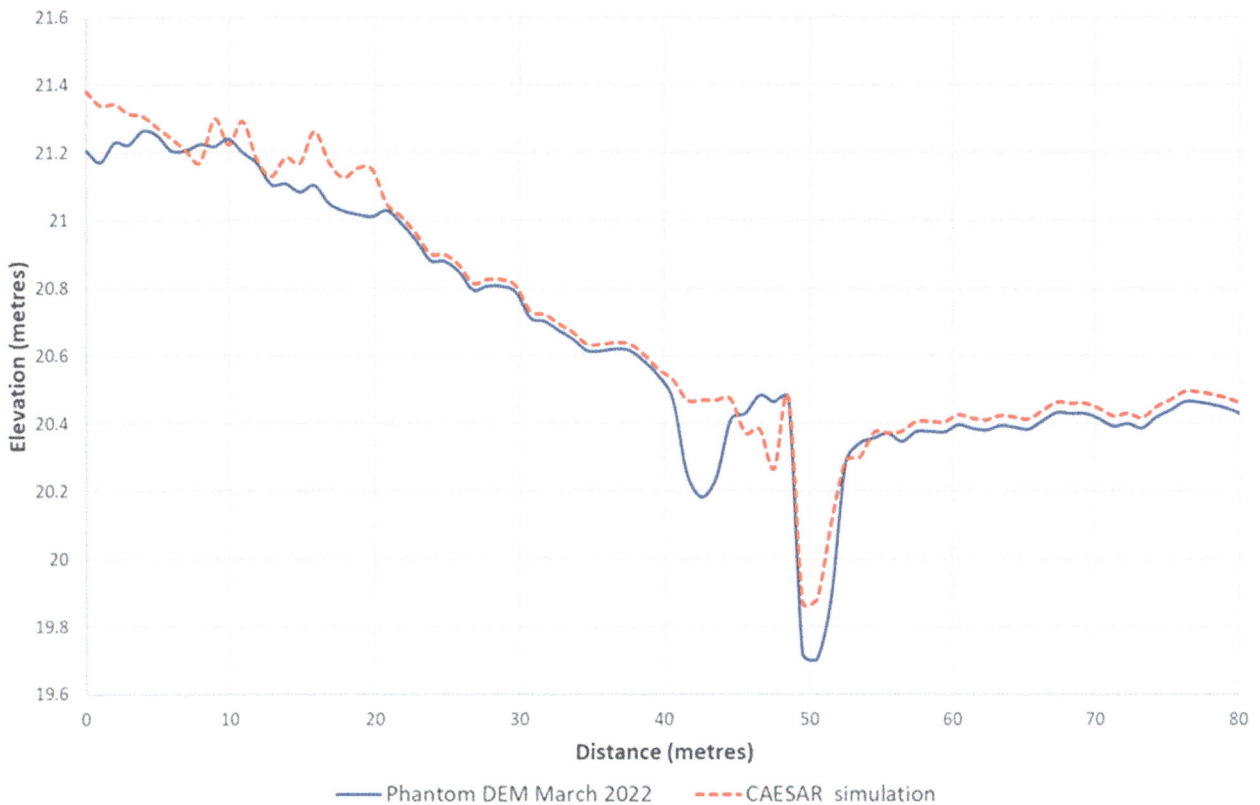

FIG 6 – Comparison of gullies predicted by CAESAR-Lisflood with those detected by RPAS flight in March 2022, with Gully 1 used as an example.

CONCLUSION

While acknowledging the study totalled only two years duration, the results here provide confidence that the parameters used in CAESAR-Lisflood are appropriate for predicting erosion at the larger catchment scale of this study. Specifically, the results here demonstrate that the model is able to correctly predict the occurrence and distribution of gullies on the Pit 1 landform.

ACKNOWLEDGEMENTS

The authors acknowledge the support and assistance of the Supervising Scientist Branch, particularly David Loewensteiner, Jay Nicholson, Hari Paramjyothi and Tim Whiteside. Energy Resources of Australia is acknowledged for their assistance with access to the study site.

REFERENCES

Hancock, G R, Saynor, M J, Lowry, J B C and Erskine, W D, 2020. How to account for particle size effects in a landscape evolution model when there is a wide range of particles sizes, *Environmental Modelling and Software,* 124:104582. https://doi.org/10.1016/j.envsoft.2019.104582

Lowry, J, Coulthard, T, Saynor, M and Hancock, G, 2020. A comparison of landform evolution model predictions with multi-year observations from a rehabilitated landform, Internal Report 663, Supervising Scientist, Darwin.

Saynor, M J, Lowry, J B C and Boyden, J M, 2019. Assessment of rip lines using CAESAR-Lisflood on a trial landform at the Ranger uranium mine, *Land Degradation and Development,* 30:504–514. https://doi.org/10.1002/ldr.3242

Wilcock, P R and Crowe, J C, 2003. Surface-based transport model for mixed-size sediment, *Journal of Hydraulic Engineering,* 129:120–128.

Alternative mineral processing strategies for copper-gold tailings management

R Narayan[1], M Edraki[2], A Scheuermann[3], M Ziemski[4], C Antonio[5] and S Quintero[6]

1. PhD Candidate, The University of Queensland, St Lucia Qld 4072.
   Email: roneel.narayan@uq.edu.au
2. Associate Professor – Group Leader – Environmental Geochemistry, The University of Queensland, St Lucia Qld 4072. Email: m.edraki@uq.edu.au
3. Professor – School of Civil Engineering, Faculty of Engineering, Architecture and Information Technology, The University of Queensland, St Lucia Qld 4072.
   Email: a.scheuermann@uq.edu.au
4. Associate Professor – Julius Kruttschnitt Mineral Research Centre (JKMRC), The University of Queensland, St Lucia Qld 4072. Email: mziemski@uq.edu.au
5. Senior Research Fellow – Julius Kruttschnitt Mineral Research Centre (JKMRC), The University of Queensland, St Lucia Qld 4072. Email: c.antonio@uq.edu.au
6. Senior Research Technologist – School of Civil Engineering, Faculty of Engineering, Architecture and Information Technology, The University of Queensland, St Lucia Qld 4072.
   Email: s.quintero@uq.edu.au

INTRODUCTION

Tailings are mostly loose material and can be classed as non-plastic sandy silts. They also tend to have low permeability, resulting in slow consolidation (settling) rates and low shear strength, even after consolidation. These qualities make them prone to liquefaction and, therefore, physical instability. In contrast, coarse waste rocks possess high shear strength and permeability when stacked in dumps. However, air and water flow-through the voids can cause acid mine drainage (AMD) with significant undesirable environmental implications. Combining the two waste streams before disposal by mixing presents a potential opportunity to harness the desirable properties of both waste streams. This practice is termed co-mingling and is prevalent within the coal processing industry. Co-mingling of metalliferous mine tailings is, however, a relatively new concept. By combining the coarse waste rocks and fine-grained tailings streams, the fine tailings tend to fill the voids between the waste rocks, reducing hydraulic conductivity and advective air (oxygen) flow into the rocks, thus possibly reducing acid mine drainage (AMD). Conversely, tailings are introduced to large frictional particles from waste rocks, providing reinforcement, stability, and possibly improved drainage.

New separation methods are being introduced to the mining sector, which have the capacity to extract waste with significantly larger particle sizes. High Voltage Pulse (HVP) is a technology in which relatively coarse particles ranging in size from 30 to 100 mm are exposed to an electric pulse discharge. This pulse preferentially fractures mineralised particles while leaving unmineralised rock (gangue) intact. The stream can then be passed via a screen, with smaller mineralised particles reporting to the undersize and removing coarser unbroken barren particles, thereby eliminating waste. Coarse particle flotation (CPF) is another emerging new technology in metalliferous mining that applies flotation separation to coarser particles than conventional flotation.

These new technologies produce different tailings (waste) streams with varying physical characteristics and possibly surface chemistries. This research aims to test new tailings mixtures by applying the co-mingling concept to blend the reject streams generated by these two technologies (HVP and CPF) with the tailings produced by conventional flotation. Combining different waste streams before disposal can improve stability and potentially reduce the acid mine drainage (AMD) rate, thus enabling early access to the tailings storage facility for rehabilitation and closure.

METHODOLOGY

The reject samples were characterised using standard and specialised laboratory tests as shown in Table 1.

TABLE 1

Different sample characterisation tests.

Characterisation	Method
Mineralogical	• Quantitative XRD • MLA • Microscopy – thin-section analysis
Physical	• PSD – wet sieving and hydrometer • Degradation test
Chemical composition	• Total digest analysis
Hydrological	• Hydraulic conductivity – constant head method • Permeability test
Geotechnical	• Specific gravity (Helium pycnometer) • Atterberg's limit (Liquid, plastic, and shrinkage limits)

The performance of the individual and blended samples was analysed using consolidation testing (ASTM D2435), strength testing (ASTM D3080–04), and permeability testing (ASTM D5084–16a) in the Geotechnical lab at the University of Queensland. In addition, the hydro-geochemical performance is currently being investigated using leachates collected from time-dependent column leaching tests. Figures 1 and 2 display a schematic and the actual experiment set-up.

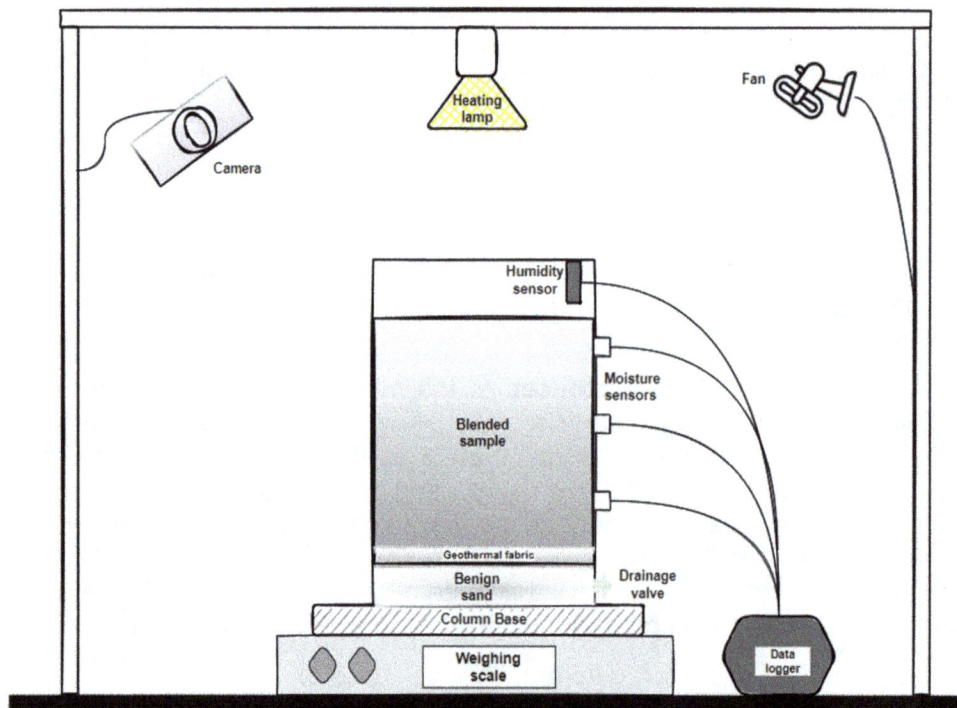

FIG 1 – Schematic of the experimental set-up.

FIG 2 – Experiment set-up.

The ongoing time-dependent column test will enhance our understanding of hydrology and geochemistry. Furthermore, this test will enable us to analyse the following:

- Compare the settlement behaviour of blended mixtures of the rejects to those of individual reject streams.

- The water balance inside the blended and individual reject streams.

- The reaction rates including AMD generation rates can be determined from the rates of solute release.

- Degradation/internal erosion – analysed from the leachate for degradation or breakdown of minerals over time.

RESULTS AND DISCUSSION

A summary of the quantitative XRD mineralogy is provided in Figure 3.

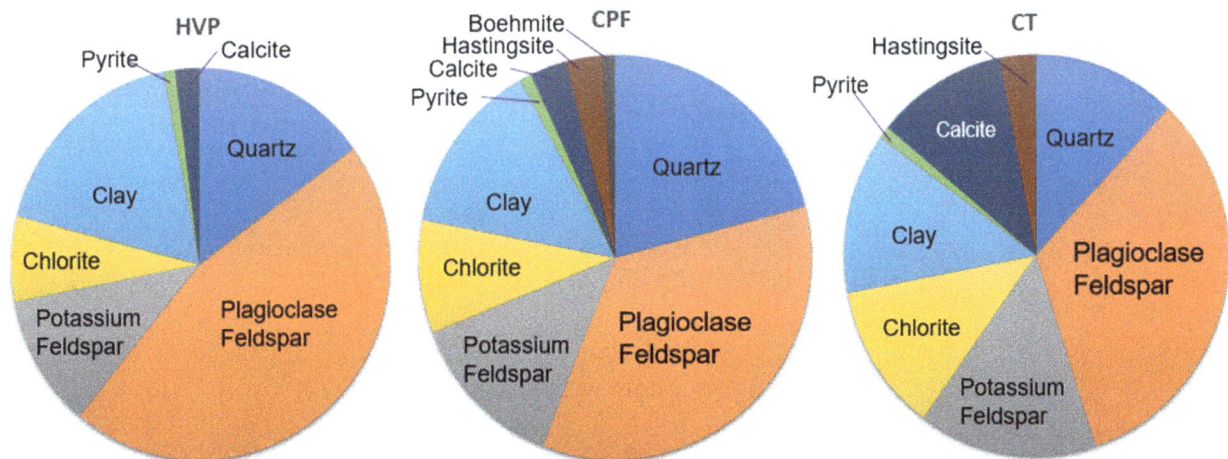

FIG 3 – Sample characterisation results – quantitative XRD.

The samples contain clay minerals (chlorite, illite/muscovite) which affect the geotechnical characteristics of tailings, such as physical degradation, and water-holding characteristics of tailings. On the other hand, plagioclase feldspar, chlorite, calcite, and hastingsite are relatively fast-weathering minerals and provide neutralising capacity to buffer acid generation from pyrite. Pyrite content accounts for approximately 1% wt of HVP, CPF and conventional tailings.

The physical performance of the reject samples were analysed using the consolidation test (Figure 4), direct shear test (Figure 5) and permeability test (Figure 6). The Compression Index (Cc) is an important parameter for analysing consolidation behaviour. The higher the Cc value, the more compressible the tested sample is.

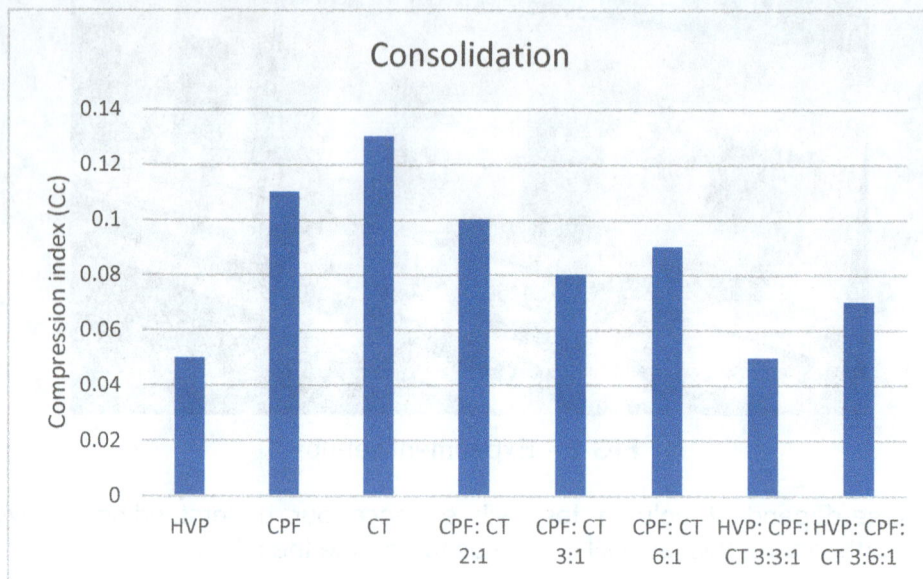

FIG 4 – Consolidation test results.

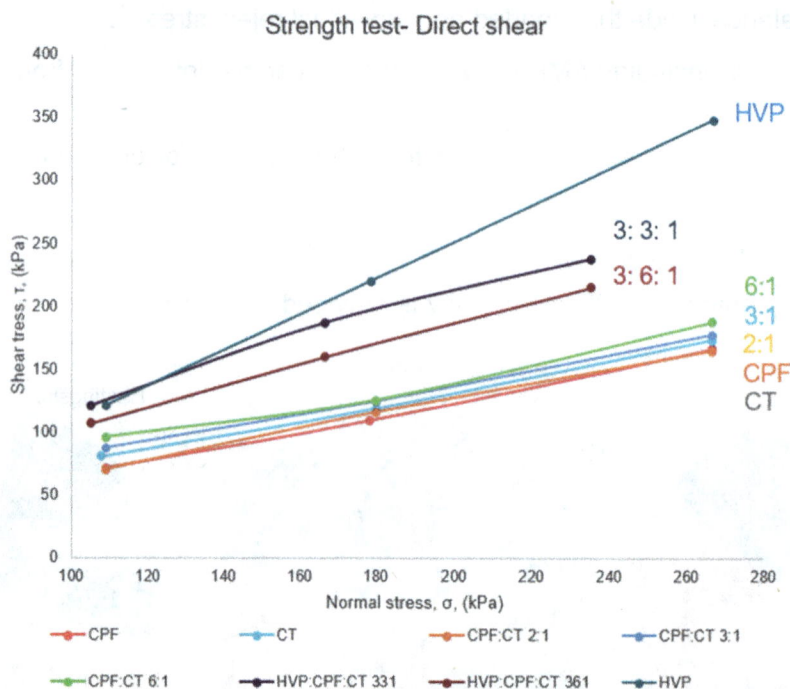

FIG 5 – Strength testing results.

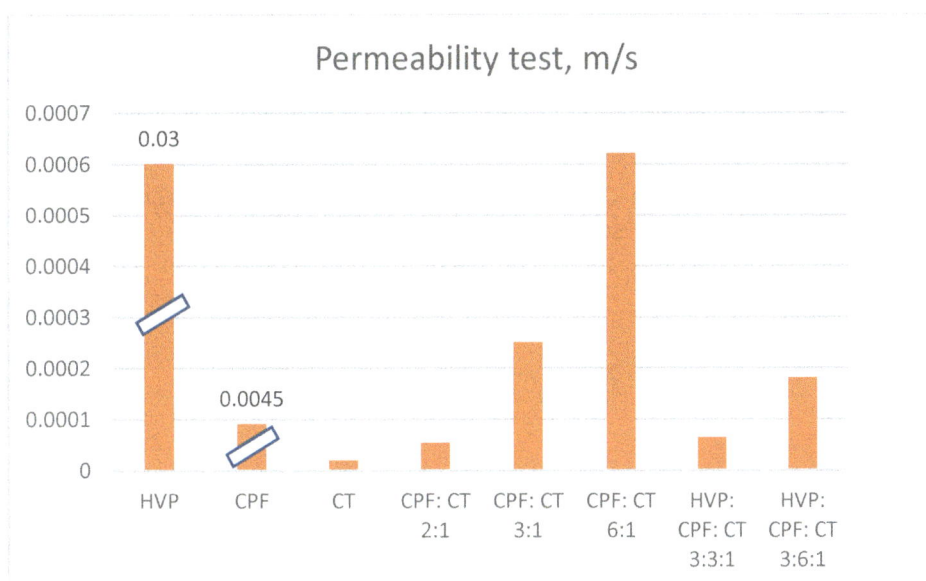

FIG 6 – Permeability test results.

Figure 4 summarises the results of the physical performance. The blends with lower size fractions (CPF and CT) displayed higher compressibility (higher Cc values) than those with larger size fractions (HVP). The CPF and CT blends showed no significant change in Cc values indicating that the compressibility is not significantly impacted by the blended mix of lower-size fractions. In contrast. Adding HVP rejects to the blend decreased the compressibility to a larger extent.

The angle of internal friction (ϕ) is an important parameter in analysing strength performance. A higher friction angle indicates stability at that angle of repose.

Strength test results indicated that the HVP rejects has the highest friction angle. The HVP rejects maintain the rock skeleton structure providing more strength and stability compared to the samples with smaller particles. The smaller individual samples (CPF, CT and CPF-CT blend mixes) show very similar friction angles indicating that the stability is not significantly impacted by the different blend mixes of smaller size fractions. The stability increases by about 10° upon adding the HVP reject to the blended mix.

The permeability test measures the rate of hydraulic conductivity (drainage) of the test sample.

The HVP sample was tested and determined to have the highest hydraulic conductivity. The CT sample was displaying the lowest hydraulic conductivity. A trend that can be extrapolated from the results was that the hydraulic conductivity increased with the proportion of CPF in the blend. The same trend can be seen with HVP samples, HVP: CPF: CT 3:3:1 blend showed a lower hydraulic conductivity value compared to the HVP: CPF: CT 3:6:1 blend.

CONCLUSION AND FUTURE WORKS

Understanding the geotechnical, hydrological, and geochemical performance of the mine residues will enable better management of the residue generated during mine operation, thereby enabling better planning and rehabilitation at mine closure. Comixing different reject types from mine operations may reduce the risk of AMD by inhibiting the ingress of oxygen and moisture to react with the sulfide minerals. It may also improve the stability of the tailings storage facility by harnessing the strength properties of rejects with larger size fractions. Comixing of different mine wastes may provide the possibility of reducing the environmental risk and provide a stable foundation for effective mine closure and rehabilitation efforts. The physical performance and geotechnical tests have been completed for this research project. The time-dependent column tests to analyse the geochemical and hydrological parameters are currently in progress.

ACKNOWLEDGEMENTS

The authors are grateful to John O'Callaghan, David Seaman, Brigitte Seaman, Dieter Engelhardt and Ian Clatworthy from Newcrest Mining Limited for their support and for providing the test samples and process data for this research project.

Dylan Carr from the Pilot Plant at Julius Kruttschnitt Mineral Research Centre (JKMRC) for assistance with experiments and test set-up.

Quantifying the influence of weathering on erosion and degradation of post mining landforms

W D D P Welivitiya[1] and G R Hancock[2]

1. Research Associate, School of Engineering, The University of Newcastle, Callaghan NSW 2308. Email: welivitiyagedon.welivitiya@newcastle.edu.au
2. Associate Professor, School of Environment and Life Sciences, The University of Newcastle, Callaghan NSW 2308. Email: greg.hancock@newcastle.edu.au

INTRODUCTION

In mining, earth is removed to access the mineral or commodity of interest. This exhumed material is placed in a waste rock dump, which is usually later reshaped to be erosionally stable, initiate and sustain vegetation and ultimately integrate with the surrounding undisturbed landscape. However, it is unclear how this exhumed material will weather and evolve for most sites. Also, there is a lack of data on the weathering of post-mining materials. If weathering is better understood, then there is the potential to improve rehabilitation outcomes. To improve predictions of the evolution of soil properties and weathering-limited erosion, estimating the rate at which fine, transportable material is produced from the weathering of larger, non-transportable particles and the resultant fragment size distribution and how that distribution evolves over time is necessary. This study examines the weathering process for four different coalmine waste rock materials for a Bowen Basin, Queensland site. Materials obtained from various areas of the coalmine were subjected to cyclic wetting and drying and heating and cooling.

Particle size distribution and other material properties were explored after each experimental cycle to ascertain how the material changes during the experimental cycles. The particle size distribution data of wet and dry cycles were used as input data in a numerical weathering model to explore the dynamics of the weathering process for each material sample. A Monte Carlo framework was employed to assess the geometry of the weathering products, weathering rate and the number of daughter particles produced within a weathering cycle. Then the SSSPAM Landscape Evolution Model was used to evaluate the influence of weathering on erosion and landscape evolution of a post mining landform. SSSPAM simulations with: (i) active weathering and (ii) without weathering were run for each material. The erosion rate and the overall landform structure were then compared between weathering and non-weathering simulations.

MATERIALS AND METHODS

A 500 g subsample (from each 20 L drum) of each material was placed into seven individual 1 L free draining plastic garden pots (Figure 1). For each material, there were seven pots. In the base of each pot, a piece of geotextile fabric was placed to ensure that only water could enter and exit the base. One pot remained at room temperature, one pot was placed in an oven at 40°C for the duration of the experiment, and the remaining five samples were subject to wetting and drying cycles.

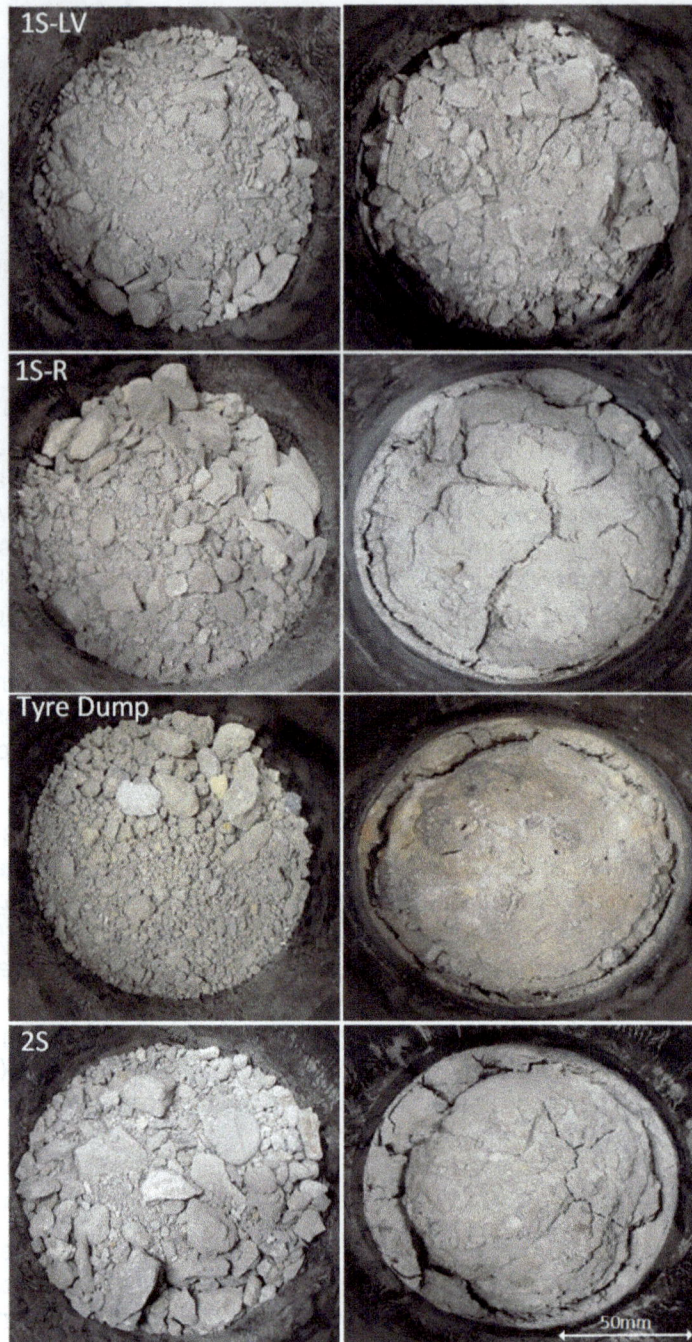

FIG 1 – Sample material at the start of the wet and dry cycles (left)
and after 20 wet and dry cycles (right).

For wetting and drying, five samples of each material were placed in a plastic tray filled to a depth of approximately 30 mm with rain water and kept wet for 12 hours (overnight). After free draining for an hour, the samples were placed in an oven at 40°C and allowed to dry for approximately five days. This wetting and drying process was repeated for 3, 6, 10 and 20 cycles. At the completion of the wet and dry cycles, the particle size distribution (PSD) of each sample was determined. Using this PSD data, the weathering module available in SSSPAM was employed to estimate the weathering rate W and weathering mechanisms of different spoils material collected from the field by fitting the module parameters to best represent experimental observations. In SSSPAM, the fracture geometry is determined using two parameters. They are the fraction of the parent particle retained by the largest daughter particle α and the number of total daughter particles n (Figure 2).

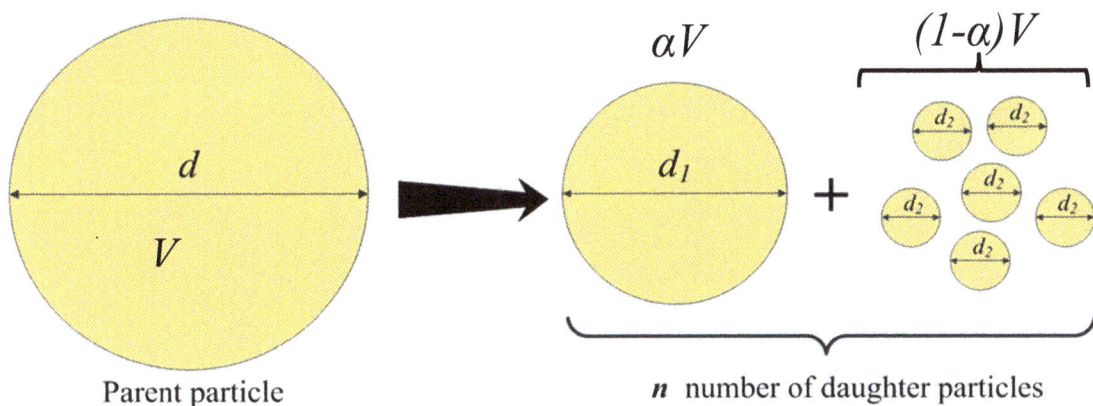

FIG 2 – Schematic diagram of weathering model used in SSSPAM.

Using the weathering parameters for different sites and previously determined erosion parameters, SSSPAM simulations were carried out to characterise the influence of weathering on the evolution of landforms constructed with different materials.

A synthetic benched landform with typical slope lengths found in post mining landforms was used in this assessment. Differences in erosion rate and the overall landform structures were examined using the non-weathering and weathering simulations.

RESULTS AND DISCUSSION

Table 1 shows the weathering parameters for different sample materials and erosion parameters. The weathering rate column (W) shows the fraction of material that will undergo weathering during a single year. It shows that although the material sampling was done in the same coalmine location, the erosion and weathering properties vary considerably from site to site.

TABLE 1

Erosion and weathering parameters for different materials.

Material site	Erosion parameters			Weathering parameters		
	α1	α2	e	W	α	n
1SLV	1.50	0.61	46.29	0.95	0.27	12
1SR	1.36	0.18	18.52	0.60	0.09	13
TD	1.08	1.80	2.82	0.39	0.07	96
2S	1.20	2.78	13.69	0.82	0.26	94

Figure 3 shows the resultant landforms after 100 years of simulated evolution with and without erosion. It shows that weathering can drastically affect how the landform evolves and the final geomorphology of the landform.

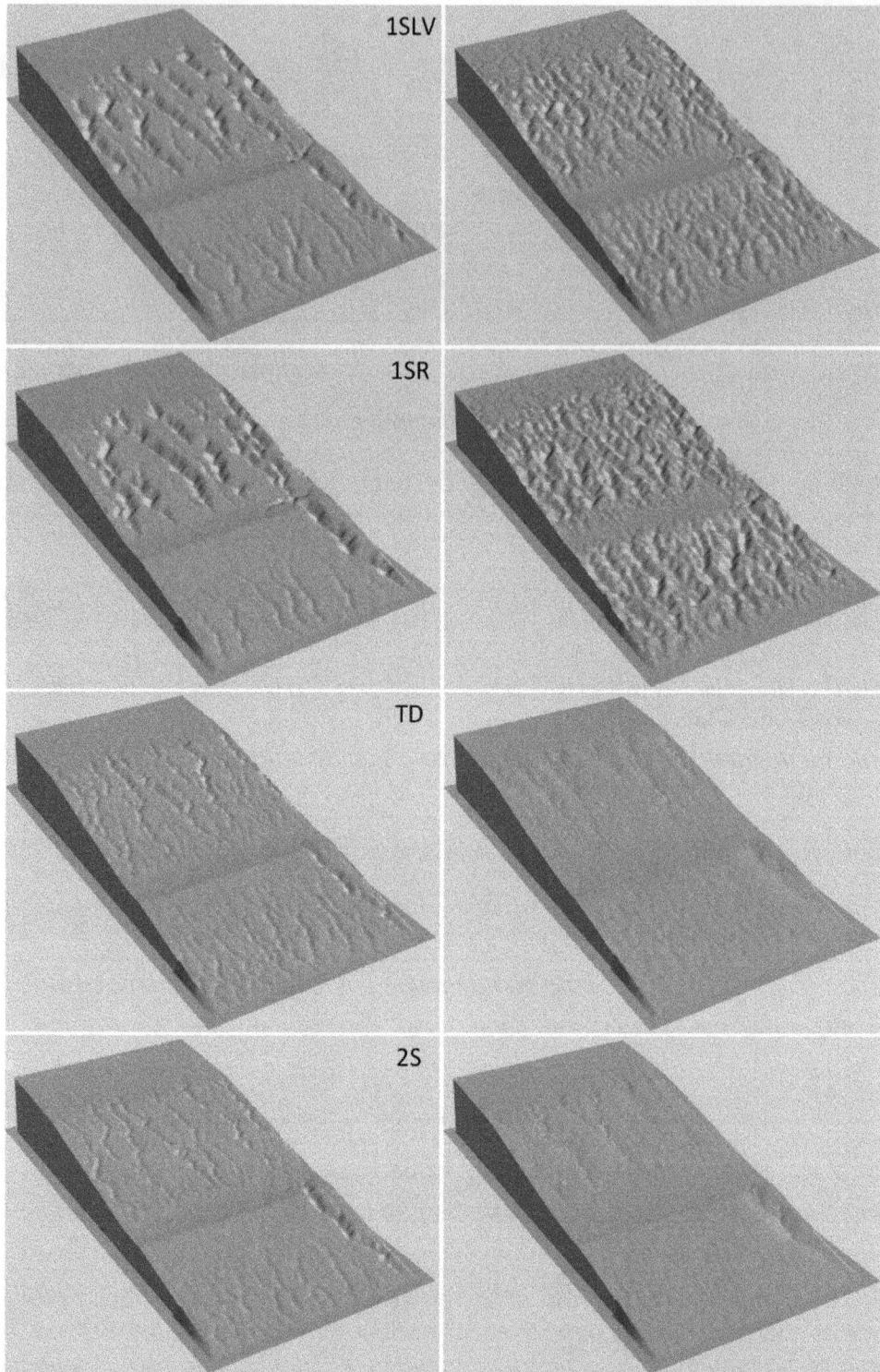

FIG 3 – Erosion characteristics of final landforms after 100 years of simulated evolution for different materials without weathering (left) and with weathering (right).

Figure 4 shows how the erosion rate of the landform changes over time for different material types with and without weathering. As evident from Figure 4, the erosion rate is significantly higher for the weathering simulations compared to non-weathering simulations across all the material types.

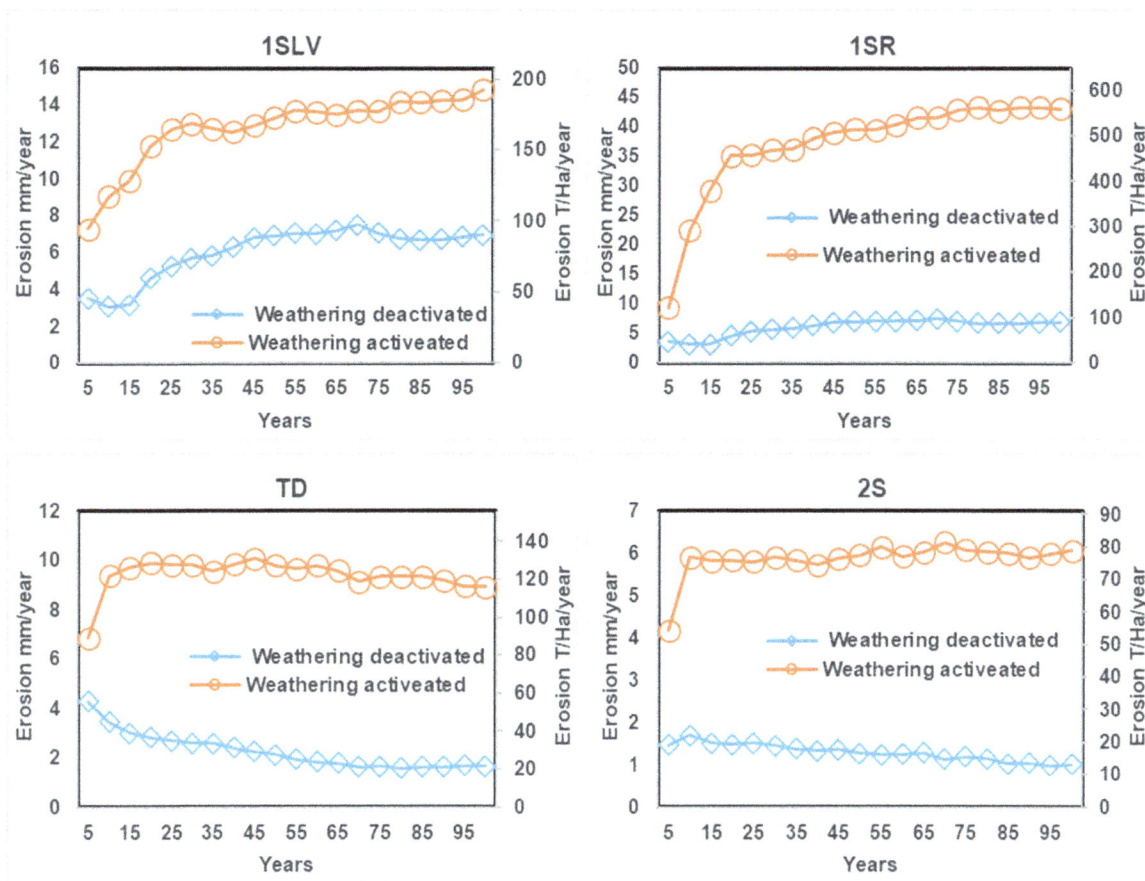

FIG 4 – Erosion rate change of the synthetic landform during 100 years of simulated evolution for different materials without weathering and with weathering.

An interesting observation from Figures 3 and 4 is that, although weathering increases the overall erosion rate, the gully incision depths are lower in weathering simulations. This can be observed in TD and 2S simulations.

CONCLUSIONS

In this study, the weathering rate and the weathering geometry of different material samples obtained from a coalmine waste rock repository were determined. The SSSPAM coupled soilscape-landform evolution model was then used to assess the influence of weathering on the evolution of landforms constructed with different materials. The results show that the weathering process could drastically increase the erosion rates and significantly influence the geomorphology of the landform. These results further reinforce the need to conduct erosion and weathering potential assessments on different materials and select suitable material types to construct (clad) future post mining landforms to minimise erosion. It also highlights SSSPAM's versatility as an analysis tool that can be used for various purposes within the mining industry.

ACKNOWLEDGEMENTS

This work was supported by the mining industry through Australian Coal Association Research Program (ACARP) Projects C34025: New Landscape Evolution Model for Assessing Rehabilitation and C27042: Adaption of design tools to better design rehabilitation and capping over highly mobile mine waste and Australian Research Council Discovery Grant DP110101216.

Integrated sustainability planning

Risks beyond relinquishment – quantifying post-relinquishment management and maintenance costs

M L Cramer[1], M R Lane[2] and N P Wines[3]

1. Director, Accent Environmental Pty Ltd, Cremorne Vic 3121.
   Email: michael.cramer@accentenvironmental.com.au
2. Director, Lane Associates Ltd, Auckland, New Zealand. Email: malcolm@laneassociates.co.nz
3. Principal Environmental Consultant, Accent Environmental Pty Ltd, Cremorne Vic 3121.
   Email: neil.wines@accentenvironmental.com.au

INTRODUCTION

Increasing attention is being focused by regulators and industry on the residual environmental liabilities associated with mining projects following site relinquishment (termed surrender in Queensland). Even sites that have been rehabilitated to a high standard and to the satisfaction of the regulator are not completely risk free and may require some degree of ongoing management. As well as ongoing monitoring and maintenance costs, considerations should include how costs associated with potential risk event occurrence can be quantified and, if considered material, how provisioning should be undertaken to minimise the risk to the state and to taxpayers.

In Queensland for example, once a resource site is fully rehabilitated and meets its environmental authority conditions or rehabilitation completion criteria, the tenure and environmental authority can be surrendered. During the surrender process any 'residual risks' resulting from the use of the site for the resource activity must be considered, and, if requested, a payment given to the government to cover any ongoing management and risk costs.

Accent Environmental, in conjunction with Lane Associates, was engaged by the Queensland Government as part of the government's financial assurance framework reforms to develop an innovative, 'one-size fits most' Residual Risk Calculator for the Queensland resources industry. The calculator is an Excel-based tool that uses detailed user inputs to estimate post-surrender management requirements and liability at resource sites.

If residual environmental risks associated with rehabilitated mine sites can be identified, quantified and managed through pragmatic regulatory tools and effective provisioning, site relinquishment/surrender can be achieved without compromising environmental and social outcomes or exposing the state to unreasonable financial risk. The effective management of residual risk can also help facilitate operations as well as relinquishment/surrender of sites where a 'walk-away' solution may not be otherwise technically feasible.

UNDERSTANDING RESIDUAL MANAGEMENT

Residual management is a term that can be used for management activities required at a former mine site – relating to the past use of the site for mining purposes – after the site has been formally relinquished/surrendered. Residual management therefore applies to activities that are required after all closure phases (including the post-closure monitoring and maintenance phase) have been completed to the extent required by the regulator and responsibility for the lease or authority has been transferred to the state or a third party (such as a landholder).

Residual management comprises two components as shown in Figure 1:

1. Ongoing management (planned monitoring and maintenance activities).

2. Residual risk management (management of risk events).

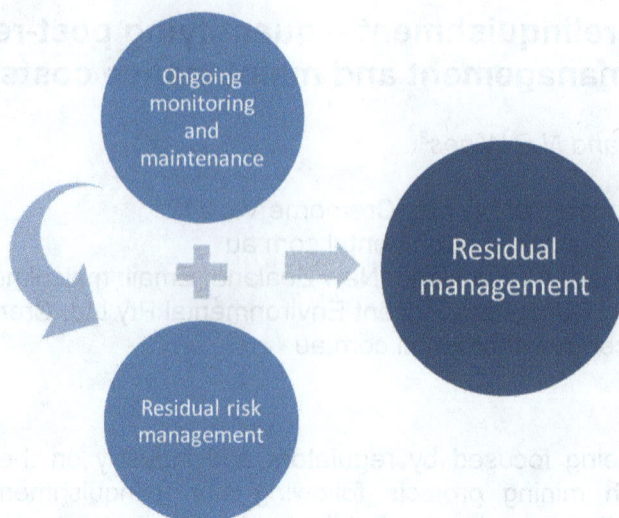

FIG 1 – Schematic of residual management.

In this context, residual risk is the risk remaining at a mine site after relinquishment/surrender that an event (risk event) may occur that requires remediation and that may trigger a need for ongoing monitoring or maintenance work – where the event is a result of past mining activities.

Sites that have met the criteria for relinquishment/surrender may still require ongoing management such as precautionary groundwater monitoring, the ongoing removal of trees from an engineered cover, or the maintenance of a fence around a pit. As indicated in Figure 1, there is also a relationship between ongoing management and residual risk, whereby ongoing management can reduce the likelihood and/or consequences of a risk event.

QUANTIFYING AND COSTING RESIDUAL MANAGEMENT

For residual management to be implemented at a relinquished/surrendered mine site, it first needs to be identified, quantified and costed. In the absence of a mechanism for considering and funding residual management, the costs of management will fall directly to the state and the taxpayer. The quantum also needs to include an appropriate level of contingent funding for events of uncertain occurrence and magnitude (ie risk events), some of which will almost inevitably occur particularly when multiple sites are considered.

The determination of ongoing management (monitoring and maintenance) requirements can be relatively straightforward. In contrast, the determination of potential credible risk events and quantification of associated remediation costs can be much more difficult. By the time a project has met the criteria necessary to achieve relinquishment, the likelihood of any single major risk event occurring should be extremely low (1 in 1000, or 1 in 10 000 years, for example). Given such infrequent occurrence, the allocation of likelihood to the events relies primarily on professional judgement as there may be little to no statistical analysis of actual risk occurrence data.

A number of approaches are available for the quantification and costing of ongoing management costs, depending on the objectives of the exercise and whether the costs are being calculated for an individual site or across an entire jurisdiction. These approaches include generic calculation tools, tailored calculation tools and expert panel processes. An expert panel process that incorporates the estimation of both closure costs and residual post-relinquishment/surrender costs has operated successfully for many years in New Zealand.

QUEENSLAND RESIDUAL RISK CALCULATOR

In Queensland, under the *Environmental Protection Act 1994*, any 'residual risks' (as termed by the Act) resulting from the use of the site for the resource activity must be considered, and, if requested, a payment given to the government to cover identified risks and any ongoing management costs. The Residual Risk Calculator has been developed by Accent Environmental and Lane Associates for the Queensland Government to assist in meeting the 'residual risk' requirements of the Act. The calculator is a generic tool designed to be broadly applicable to the Queensland resources industry, including mining and petroleum and gas projects.

The calculator estimates post-surrender liability based on standard risk assessment principles, consideration of 18 credible post-surrender risk events and determination of likely post-surrender management requirements. The tool provides a simple-to-use interface that feeds site data into a series of algorithms which generate site risk profiles and management schedules, and estimate associated post-surrender costs. The architecture of the calculator is shown in Figure 2.

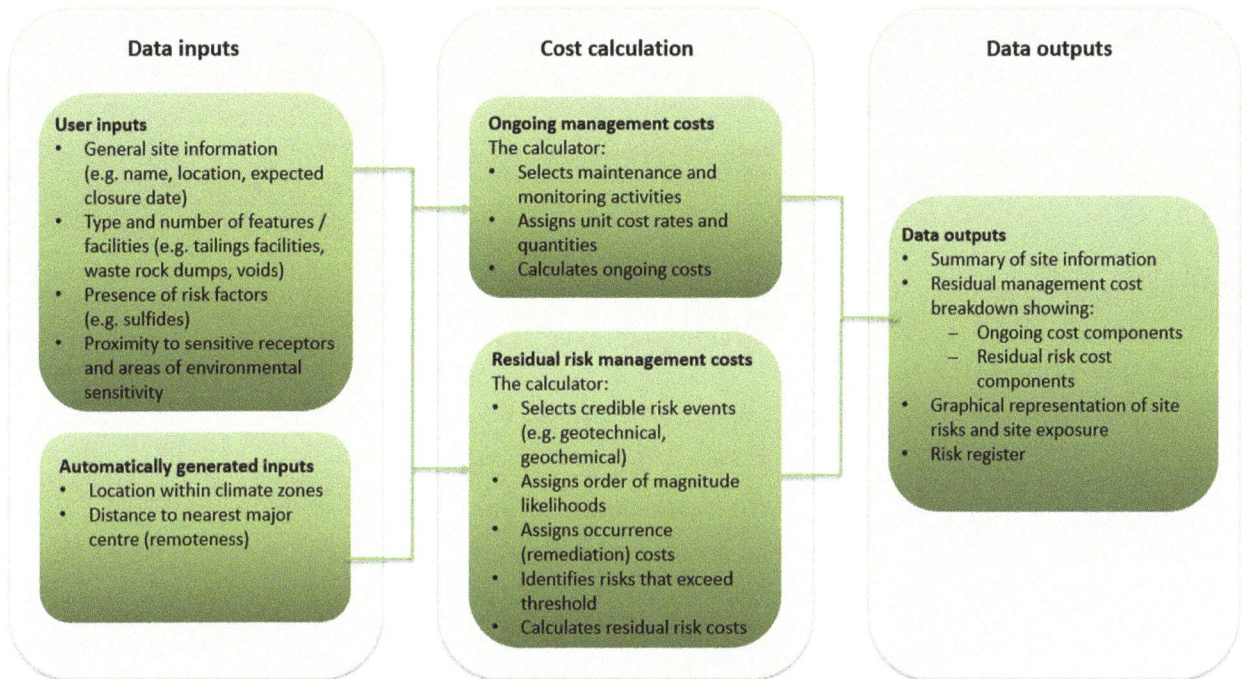

FIG 2 – Architecture of Residual Risk Calculator.

The calculator complements a proposed expert panel process that will enable consideration of more site-specific factors in determining residual management costs. It also complements the use of Queensland's estimated rehabilitation cost (ERC) calculators which are used to calculate the ERC during operation and closure.

The Residual Risk Calculator has been developed to meet the needs of the Queensland resources industry and financial assurance framework reform. However, the principles underpinning the tool have wide application. There is great potential for other jurisdictions and for industry to adapt the Queensland approach to understand and manage the issue of residual management at mine sites and thereby to further environmental and social outcomes and protect the state or company from unreasonable and avoidable financial risk.

ACKNOWLEDGEMENTS

Accent Environmental and Lane Associates would like to thank the project teams at the Queensland Department of Environment and Science and Queensland Treasury Corporation for their support and inputs in developing the Queensland Residual Risk Calculator.

Whole of mine planning – Evolution Mount Rawdon case study

A Ferrier[1], A Forbes[2], G Maddocks[3] and M Landers[4]

1. Product Manager – CAD and Enviro, Deswik, Brisbane Qld 4000.
   Email: ainsley.ferrier@deswik.com
2. Principal Mining Consultant, Deswik, Brisbane Qld 4000. Email: amanda.forbes@deswik.com
3. Principal Hydrogeochemical Engineer, RGS Environmental Consultants Pty Ltd, Coopers Plains
   Qld 4108. Email: greg@rgsenv.com
4. Principal Geochemist, RGS Environmental Consultants Pty Ltd, Coopers Plains Qld 4108.
   Email: matt@rgsenv.com

INTRODUCTION

Evolution's Mount Rawdon Operation (MRO) produces gold and silver through open pit mining methods. The mining operations are scheduled to be completed in FY24 (Financial Year 2024), after which low-grade stockpiles will be processed until the end of FY27.

As part of MRO's Progressive Rehabilitation and Closure Plan (PRCP), RGS Environmental Consultants (RGS) and Deswik were engaged to develop the final landform surface and a whole of site probabilistic water balance/water quality (WB/WQ) model to evaluate the viability of several closure options with respect to water quality in the various site water storages, particularly the final void which will form a pit lake. This case study details the closure option which included (amongst other things) covering the final landforms and directing water toward or away from the final void post closure.

Closure and rehabilitation for MRO had previously been planned and evaluated on a 'domain' basis. Furthermore, the rehabilitation plans and final water management plans for the Tailings Storage Facility (TSF), Waste Rock Storage Facility (WRSF) and other domains were not coordinated, possibly resulting in competitive rehabilitation material requirements at the commencement of closure activities. An assumption was that a legacy non-acid forming (NAF) stockpile would provide sufficient material to both supplement ex-pit material for the construction of the TSF embankment raises and to complete all closure activities, but the volumes available and the volumes required to complete the closure activities had not been quantified.

METHOD

A Whole of Mine (WOM) landform model was developed by RGS and Deswik to sequence the remaining mining and potential closure activities into a single visualisation to assess the viability of the selected closure option, quantify the overall rehabilitation material balance and optimise material placement to reduce costs and improve closure outcomes.

To develop the WOM model, Deswik modified MRO's existing operational landform and haulage model to practically align and sequence the landform designs for the WRSF, ore stockpiles and TSF with the remaining mined material classifications and volumes. This provided:

- A spatial and time-based representation of a practical mining and dumping sequence for the remainder of the pit and processing operations.

- Assurance that potentially acid forming (PAF) material could be practically placed and encapsulated within the WRSF during operations to mitigate acid and metalliferous drainage (AMD) potential.

- The quantity of 'useful' material mined during pit operations.

- The quantities and overall material balance at the commencement of closure activities.

- The spatial profile of the low-grade ore stockpiles (which impact specific waste placement options on the WRSF) at the completion of pit operations.

- The spatial profile of the WRSF after reclaiming the long-term ore stockpiles.

- The spatial profile of the TSF at the completion of processing.

Conceptual cover systems for the WRSF were developed based on the potential rehabilitation materials available and their material properties. The conceptual cover designs were incorporated into the WOM model, which showed a significant material deficit for the site. The volume within the legacy NAF stockpile was insufficient for closure activities, and borrow areas were required to meet the volume requirements to construct the TSF cover system over the final tailings beach to drain surface flow from the TSF perimeter to the external drop structures. High-level borrow areas were added into the WOM model to address this issue.

RESULTS

The spatial and time-based sequencing was reviewed to identify and incorporate material placement opportunities and resolve risks for both mining operations and closure activities, including:

- Placing ex-pit PAF within the TSF to reduce rehandle requirements to construct the TSF cover system after mining had cease and reduce adverse environmental risk with rehabilitation on the WRSF.

- Rehandling PAF on the WRSF in the next financial year to ensure sufficient capacity for the low-grade ore stockpiles.

RGS developed the whole of site probabilistic (rainfall) WB/WQ model using the resultant WOM model to provide operational, closure and post closure inputs, including the optimised final landform surface, catchment areas and staged curves. The WB/WQ model was constructed using the mass balance approach with GoldSim software and the Contaminant Transport (CT) module, and was used to:

- predict water quality in the various site water storages (including the final void/pit lake)

- model how the water qualities will evolve 100-years into the future

- model the demand for water to fill and potentially overflow from the pit lake during and beyond closure

- assess how climate change and climate variability are likely to affect the water balance and water quality of the site water storages, specifically the pit lake, using deterministic and stochastic (probabilistic) climate sequences.

The WOM model was integrated with the water modelling to verify catchment areas that report to each node in the GoldSim model and enabled RGS to evaluate the effect of the material placement strategy on water quality in the various water storages particularly the pit lake.

DISCUSSION

This case study is an outstanding example of designing for closure and implementing integrated planning. The work brought together the traditionally siloed mine planning, tailings, and environmental teams to deliver a Whole of Mine integrated mining and closure landform and haulage model (WOM model) as part of their Progressive Rehabilitation and Closure Planning (PRCP).

Design and calibration of a climate resilient Type B (adapted) sediment basin at the Dugald River Mine in the north-west Queensland minerals province

G Green[1], A Harburg[2] and A Dendys[3]

1. Environment Superintendent, MMG Dugald River Mine, Cloncurry Qld 4824.
   Email: gemma.green@mmg.com
2. Environmental Advisor, MMG Dugald River Mine, Cloncurry Qld 4824.
   Email: amy.harburg@mmg.com
3. Environmental Advisor, MMG Dugald River Mine, Cloncurry Qld 4824.
   Email: amanda.dendys@mmg.com

INTRODUCTION

Dugald River Mine (DRM) is located approximately 85 km north-east of Mount Isa, and 65 km north-west of Cloncurry, in the north-west Queensland Minerals Province of Australia. DRM is an underground operation and includes an above ground processing area, a 200 ha tailings facility, two waste rock dumps and associated infrastructure. The mining method is conventional underground longhole open stoping and downhole benching of a lead/zinc deposit within a black slate environment. The deposit is estimated to contain ~53 Mt of zinc and lead, which will be mined over a 30-year period.

During operations, waste rock is segregated into either Potentially Acid Forming (PAF) or Non-Acid Forming (NAF) based primarily on the sulfide concentration of the rock. In particular, the NAF waste rock is an asset and is used for various small construction projects during operation; however, its primary use is for rehabilitation work at closure. Since beginning operations in 2017, DRM has generated more NAF than was originally anticipated. Whilst this is an excellent asset to DRM, it did pose a logistical problem of where to store the additional NAF for the life-of-mine and how to manage the run-off.

DRM experiences distinct wet and dry seasons with hot conditions and periods of rainfall between November to April, and relatively dry and mild conditions between May to October. Majority of rainfall is received between January and February with an annual average rainfall of 501.1 mm. Short and intense rainfall events are common. DRM's above ground infrastructure is land-locked between two large ephemeral watercourses, making the significant expansion of any assets challenging from a permitting/regulation and cost perspective. MMG is also a member of the International Council on Mining and Metals (ICMM) and is committed to adapting to a changing climate through operations and closure.

Following extensive studies DRM opted to store the additional NAF material into the adjoining NAF Dam and concurrently upgrade a previously minor sediment basin into a highly engineered sediment basin which relies on automatic flocculant dosing to meet the required discharge quality limits. This basin is climate-resilient in that the dosing system can be adjusted to counter high intensity extended rainfall events which produce highly variable inflow water quality. This abstract broadly describes how DRM designed and implemented an engineering innovation to enable the site to meet the changing climatic challenges. The abstract then describes the process to design the sediment basin and how the dosing rate was established and validated. Finally, the abstract comments on the effectiveness of the sediment basin and how it has revolutionised DRM's approach to water treatment and management; all whilst minimising the disturbance footprint of the operation.

METHODS/APPROACH

The design and construction of the upgraded sediment basin needed to achieve the following outcomes:

- Achieve compliance with DRM's Environmental Authority by maintaining a 50 m buffer to a nearby watercourse which was already only 53 m from the current sediment basin.

- Achieve the required Release Water Quality Limits (as outlined in Table 1).

- Allow sufficient time for water quality samples to be taken and analysed prior to a release.

- Incorporate MMG's commitment through ICMM to climate resilience through operations.

TABLE 1

DRM's release water quality limits.

Quality characteristic	Release water quality limit
Electrical conductivity	1000 µS/cm
Aluminium (total)	5 mg/L
Arsenic (total)	0.5 mg/L
Sulfate (total)	1000 mg/L
Cadmium (total)	0.01 mg/L
Copper (total)	1 mg/L
Lead (total)	0.1 mg/L
Nickel (total)	1 mg/L
Zinc (total)	20 mg/L

The IECA (International Erosion Control Association, Australasia, 2008) defines four types of sediment basins which are summarised as:

- **Type A**: Designed to handle high flow rates and remove settled sediment via a decant structure.

- **Type B**: Similar to Type A, but without a decant structure. Type B basins can be used in conjunction with a Type A basin to help remove sediment from water that is flowing too quickly for a Type A basin.

- **Type C**: These basins are designed to store large volumes of sediment-laden water and are suited to coarse grained soils.

- **Type D**: These basins are designed to capture, treat, and release sediment laden water and are typically deep and narrow.

A Type B basin was selected noting:

- A Type A basin requires a floating decant system and based on the topography of the area this would have been below ground level and not viable.

- The soils within the catchment did not meet the criteria for a Type C.

- The effectiveness of Type D basins has been questionable, and this risk was unacceptable to DRM.

The process to refine the design included test pits, soil tests and establishing dosing rates prior to an overall risk assessment against the original design outcomes.

Test pits

Given the limited footprint, and the mineralised nature of the environment, it was important to first develop test pits to establish if/how the material could be excavated to the required depth and if any contaminants were present. To achieve this, three geographically dispersed tests pits were excavated within the dam to a depth of 3.5 m. The test pits revealed the soil profile, or horizons, and samples were collected from each level for further mechanical and chemical analysis. An additional test pit was excavated outside the proposed footprint in the event that the Type B basin was not favourable, and a larger footprint was required.

Establish dosing rate

The dosing rate is based on:

- the flow rate of the stormwater run-off
- the concentration of sediment in the stormwater run-off
- the desired level of sediment removal
- the type of flocculant being used.

The next step was to determine an appropriate dosing rate based on the sediments to be mobilised in the incoming water. As no incoming water was available, a series of soil samples were collected from drainage lines within the catchment of the sediment basin, and from the floor of the existing sediment basin. Jar tests and soil characterisation were then used to determine an appropriate flocculant and dosing rate. Based on historical water quality data, and data from the sediment collection, it was established that Aluminium Chlorohydrate (ACH) was the proposed flocculant.

Using Jar Tests, the soil samples were then mixed at a rate of 10 grams (sifted) soil per litre of water and dosing rates of zero (control), 0.01 ml/L, 0.03 ml/L, 0.06 ml/L, 0.08 ml/L and 0.1 ml/L were then applied. The treated water from the Jar Tests were then further analysed in a NATA accredited laboratory to allow comparison against DRM's discharge limits. The Jar Tests concluded that a dosing rate of 0.6 ml/L was sufficient to treat the water to the required standard after 15 minutes (see Figure 1).

FIG 1 – Jar Test (0.6 ml/L on the left, Control on the right).

Refine design

One of the design outcomes was the ability to sample and analyse a water quality result prior to discharge under normal operating conditions. For a Type B basin, if the minimum storage depth were constructed, the capacity (5.5 ML) would be insufficient to contain run-off for a 24-hour ARI 1:1 event. However, with the storage excavated to 3.5 m below the spillway, the basin would have the storage capacity of 10.5 ML which would allow for the containment of 168 hour (5 day) ARI 1:1 event (provided storage was empty at onset of rainfall).

Greater retention time provides many benefits, namely:

- increased retention time allows for water quality testing
- redundancy, if the flocculant dosing is ineffective, water could be pumped to another structure.

For these reasons, DRM opted for an adaption to the standard Type B design and increased the depth from 1.5 m to 3.5 m.

Design summary

DRM's Type B sediment basin consists of three zones. The first zone known as the Inlet Zone is a rock lined structure with a trapezoidal concrete weir that houses the flocculant dosing sensor and delivery injector before the stormwater enters zone two, known as the Forebay Mixing Zone. The residence time in the Forebay Mixing Zone is short and is primarily designed as a mixing pond where primary coarse sediment deposition occurs. The treated water then flows over a second concrete weir into the Settling Zone. In the Settling Zone further deposition of finer suspended solids occurs and velocity is reduced prior to release. The Settling Zone has a rock lined spillway with a rectangular concrete weir which reports to the receiving environment. The outlet spillway is the approved Environmental Authority release location and during releases is monitored for volume via telemetry and grab samples are collected for quality, see Figure 2.

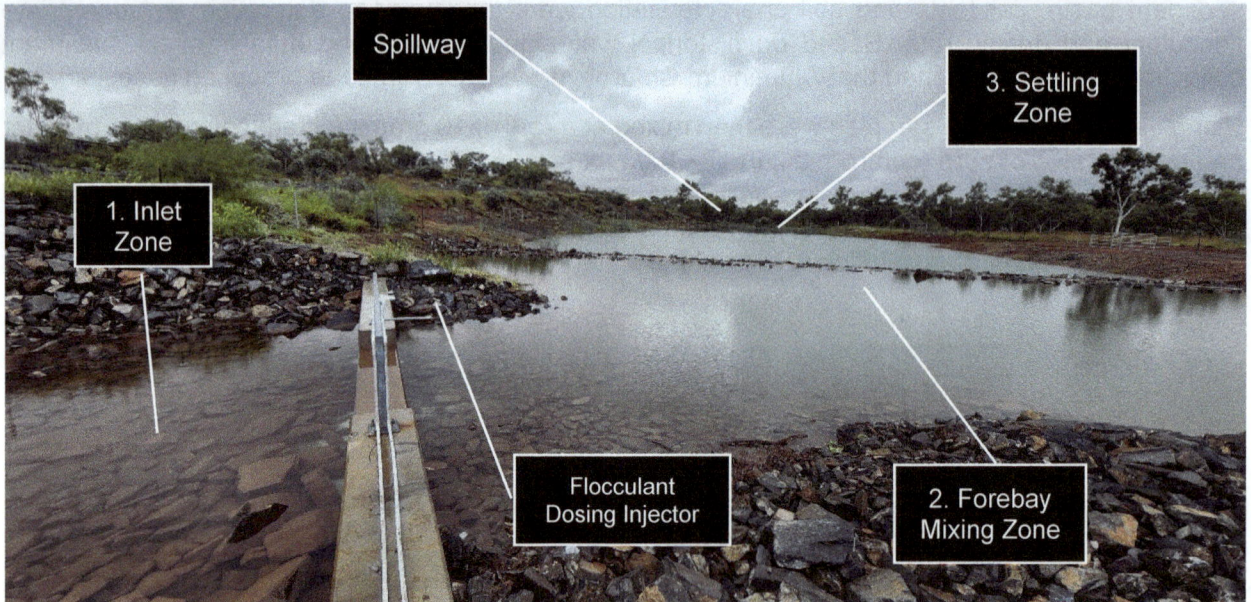

FIG 2 – Adapted-Type B sediment basin at Dugald River Mine (18/02/2023).

RESULTS AND DISCUSSION

Construction of the upgraded sediment basin was complete in November 2022, in preparation for the 2022/2023 wet season. To ensure appropriate calibration of the Sediment Basin MMG collected water samples from the Inlet Zone during any inflow, and from the Forebay Mixing Zone and Settling Zone for a period of five days following each inflow event.

Between 1 November 2022 and 31 May 2023 there were a total of 13 inflow event days. Of these, eight days were considered to be under normal conditions. Refer to Table 2 for averages of these events.

TABLE 2

Average inflow and outflow from the sediment basin under normal operating conditions.

Quality characteristic	Inlet zone	Spillway	% Reduction
Aluminium (total)	1.230 mg/L	0.321 mg/L	74%
Arsenic (total)	0.002 mg/L	0.001 mg/L	50%
Cadmium (total)	0.007 mg/L	0.000 mg/L	100%
Copper (total)	0.005 mg/L	0.001 mg/L	80%
Lead (total)	0.014 mg/L	0.001 mg/L	93%
Nickel (total)	0.001 mg/L	0.000 mg/L	100%
Zinc (total)	0.091 mg/L	0.005 mg/L	95%
Electrical conductivity	1960.000 µS/cm	327.000 µS/cm	83%
Sulfate (total)	981.625 mg/L	91.750 mg/L	91%

The results at the Spillway were below those required by DRM's Environmental Authority and broadly align with that expected by the Jar Tests. It is important to note that ACH showed no reduction in Cadmium of Nickel; however, as the incoming water was at the Limit of Reporting of the laboratory, it may simply indicate that there was no contaminant present to drop out of solution.

For the remaining five inflow events, DRM experienced an extreme rainfall event which altered the catchment of the sediment basin and by extension the quantity and quality of the incoming water changed. Whilst this was not ideal, it did provide insight into how the upgraded sediment basin can perform should the incoming water quantity and quality change without adjusting the dosing rate (refer to Table 3).

TABLE 3

Average inflow and outflow from the sediment basin under abnormal operating conditions.

Quality characteristic	Inlet zone	Settling zone	% Reduction
Aluminium (total)	0.082 mg/L	0.176 mg/L	-115%
Arsenic (total)	0.000 mg/L	0.000 mg/L	0%
Cadmium (total)	0.017 mg/L	0.029 mg/L	-71%
Copper (total)	0.001 mg/L	0.001 mg/L	0%
Lead (total)	0.014 mg/L	0.010 mg/L	29%
Nickel (total)	0.008 mg/L	0.013 mg/L	63%
Zinc (total)	18.062 mg/L	25.180 mg/L	-40%
Electrical conductivity	4112.000 µS/cm	2244.000 µS/cm	44%
Sulfate (total)	2288.000 mg/L	1182.000 mg/L	49%

In comparison to the normal operating conditions, it can be concluded that the dosing rate was not sufficient to treat the quantity or quality of the incoming water. It is noted that Aluminium, Cadmium and Zinc also increased in concentration and the initial investigation suggests that the metals which had previously dropped out of solution had re-mobilised during the rainfall event. Despite this, Lead, Nickel, Electrical Conductivity and Sulfate all continued to reduce. There was no change in Arsenic concentrations. MMG continues to investigate this matter.

CONCLUSION

The best type of sediment basin for a particular area and/or industry will depend on the amount of sediment, the type of sediment, the incoming flow rate, the desired water quality, and the budget available for both construction and maintenance. Furthermore, traditional sediment basin designs are based on known climatic conditions. In a world of climate change, it is vital that operators consider not only the current conditions, but how these conditions may adapt to a climate-uncertain future and by extension how an operation may continue to meet its environmental commitments in protecting the receiving environment.

The adapted Type B Sediment Basin was able to be incorporated into the existing stormwater management system at DRM without significantly altering the existing stormwater structures within the catchment. However, as the basin requires flocculant dosing, an adequate supply of flocculant is required to be kept on hand. In addition, annual maintenance of the sediment basin and the dosing system is pivotal in its success. It is also acknowledged that the calibration of the dosing rate will be an ongoing commitment even when the expected water quality is considered a known-known.

In closing, it is considered that the adapted Type B Sediment Basin is a success at DRM. It is further considered that the use of flocculant dosing is a valuable tool for stormwater management to improve the quality of stormwater run-off and protect the environmental values of the receiving environment.

ACKNOWLEDGEMENTS

Mr Andrew (Drew) Holzheimer; MMG Dugald River Mine; ATC Williams; Turbid Water Solutions; Wulguru Technical Services; Department of Environment and Science; and the anonymous reviewer of this extended abstract.

REFERENCES

International Erosion Control Association, Australasia (IECA), 2008. *Best Practice Erosion and Sediment Control (BPESC) document* [online]. Available at: https://www.austieca.com.au/publications/best-practice-erosion-and-sediment-control-bpesc-document [Accessed 10 June 2023].

Management of riverine disposal of tailings and waste rock for AMD control at the Ok Tedi copper-gold mine

R Schumann[1], W Stewart[2], S Davidge[3], B Rathi[4], E Kepe[5], J Pile[6] and M Ridd[7]

1. Principal Environmental Chemist, Environmental Geochemistry International, Balmain NSW 2041. Email: russell.schumann@geochemistry.com.au
2. Managing Director, Environmental Geochemistry International, Balmain NSW 2041. Email: warwick.stewart@geochemistry.com.au
3. Principal Hydrogeologist, Environmental Geochemistry International, Balmain NSW 2041. Email: shaun.davidge@geochemistry.com.au
4. Senior Hydrogeochemical Modeller, Environmental Geochemistry International, Balmain NSW 2041. Email: bhasker.rathi@geochemistry.com.au
5. Manager Environment, Ok Tedi Mining Limited, Tabubil, Western Province PNG. Email: erizo.kepe@oktedi.com
6. Deputy General Manager, Ok Tedi Mining Limited, Tabubil, Western Province PNG. Email: jesse.pile@oktedi.com
7. Aquatic Geochemistry, Ravenshoe Qld 4888. Email: michaelj.ridd@gmail.com

INTRODUCTION

Geographical constraints, including steep terrain and high rainfall, at the Ok Tedi copper-gold mine in PNG's Western Province present Ok Tedi Mining Limited (OTML) with significant challenges in mine waste management. Waste management includes discharge of tailings into the local river system and storage of waste rock from the mine pit in waste rock dumps which erode into the river system. Copper bearing skarn mineralisation is associated with a massive sulfide (mainly pyrite) deposit. Consequently, waste rock and tailings can contain significant amounts of pyrite. Without appropriate management, these waste streams represent a risk of producing acid and metalliferous drainage (AMD) when discharged to the river system. Throughout the life of the project, OTML have developed evolving strategies to manage mine wastes and mitigate AMD.

AMD CONTROL AT OK TEDI

To moderate the risk of AMD, OTML have developed an innovative and integrated system to manage tailings and waste rock disposal. Planning for mine waste management begins with the annual Strategic Planning cycle that develops a revised Life-of-mine (LoM) plan annually. Multiple possible LoM plans under various scenarios are generated using an optimisation scheduling program (COMET) which includes the environmental constraints (geochemical and sedimentological) that apply to the operation. Thus, all potential LoM plans (which may number in the hundreds) considered in each year, satisfy those constraints. Environmental models (geochemical and sedimentological) are run on a subset of the possible LoM plans and a single preferred LoM plan is selected that is considered to be the optimal balance of economic, social and environmental costs and benefits.

OTML uses an AMD block model to manage materials handling at the mining stage. Development of the model starts at exploration and continues through resource development, grade control and mining, with Total S assayed on drill core samples through each stage, including routine monitoring of blasthole samples. This provides the AMD model with the acid generating component of the acid-base account (ABA) for mined materials. OTML recently completed an extensive study into the acid neutralising capacity (ANC) of each of the mined rock types, and data from this study provides the acid neutralising component of the ABA assessment to the updated AMD block model. As a part of this study, alternative methods to quantify ANC were investigated and total carbon (TC) analysis was identified as the most appropriate method. The use of results from routine analysis of TC in the AMD block model is currently being implemented.

OTML are licenced to deposit waste rock in failing waste rock dumps which erode into the river if they ensure deposited rock ABA achieves a quarterly average excess ANC of 150 kg H_2SO_4/t. To achieve this target limestone is mixed with waste rock, with the AMD block model used for mine scheduling to ensure the ABA target is achieved.

To ensure AMD control, OTML's environment permit has set a limit for an average excess ANC of 30 kg H_2SO_4/t for tailings discharged into the river. This is accomplished using a multi-pronged approach. After recovery of copper and gold via flotation, tailings are sent to the tailing pyrite processing (TPP) plant where pyrite is recovered and transported to a pit through a 126 km long pipeline where it is permanently stored sub-aqueously. The final (desulfurised) tails from the TPP, which typically contain less than 1 per cent sulfur, are discharged into a tributary of the Ok Tedi. To ensure the final tailings contain excess ANC, limestone is added to ore feed at the plant mill. To achieve the target excess ANC without compromising metallurgical performance, OTML has developed a procedure to forecast the tailings mass, sulfur content and ANC and therefore excess ANC values for at least two weeks beyond the current operational date. This procedure uses a data driven modelling approach which combines statistical techniques with machine learning principles (Rathi *et al*, 2022). The forecasted tailings outputs are used to derive a daily schedule for additional limestone dosing required in the ore feed to meet the quarterly excess ANC target for tailings.

Desulfurisation of tailings prior to discharge, with the recovered sulfides deposited sub-aqueously, has allowed controlled deposition of mine waste in stockpiles on the floodplains downstream from the operation to reduce the river sediment load. Tailings are transported by fluvial action for approximately 110 km downstream to Bige where they are dredged from the river channel and stored in engineered stockpiles on the riverbank. Dredged sediment is analysed to determine its AMD properties (ANC and maximum potential acidity (MPA)) with the ANC/MPA ratio used to manage deposition.

One of these stockpiles is now undergoing remediation for closure, which has involved capping with alkaline materials, revegetation and water drainage management. OTML is currently undertaking an extensive study to investigate AMD loads from the stockpile and validate the closure strategy. Measurement of ANC/MPA ratios of samples obtained from the east bank stockpile demonstrate that material with ANC/MPA ratios below 1 (higher risk of AMD) or between 1 and 2 (moderate risk of AMD) occur only in the phreatic zone, with samples from the vadose zone having ANC/MPA ratios above 2 (low risk of AMD) (Figure 1). These results suggest that the deposition strategy of placing higher risk material in saturated regions of the stockpile and only low risk materials in the regions where saturation is not permanent has been successful and as a consequence the risk of AMD from the stockpile greatly reduced.

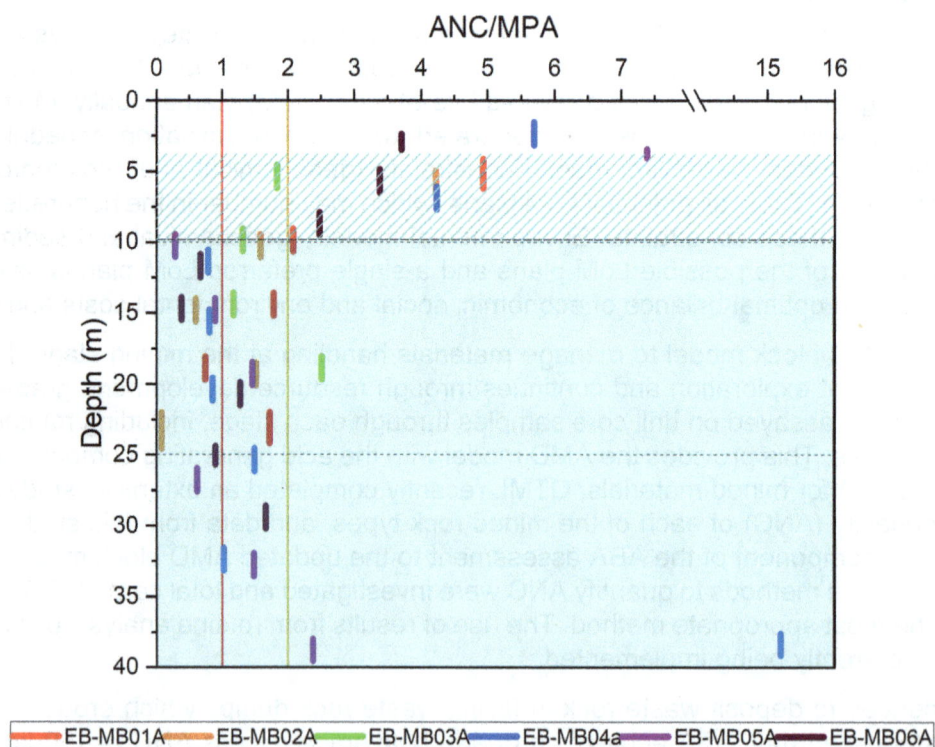

FIG 1 – ANC/MPA ratio for dredged river sediment in the east bank stockpile at Bige. The region shaded blue shows the approximate depth of permanent saturation, while the area with the hatched blue shading shows the variation in depth to the water table.

EFFECTS OF AMD MANAGEMENT ON RIVER WATER QUALITY

The combined effects of OTML's evolving mine waste management strategy, including refinement of the AMD block model for mine waste scheduling, tailings desulfurisation, optimisation of limestone addition to tailings and management of dredged river sediment stockpiles has had a positive impact on river water quality. Figure 2 shows the temporal trend in dissolved copper concentrations in the Ok Tedi between 2004 and 2021. Copper concentrations have declined during this period as a result of improved mine waste management, resulting in reduced ecological risks (Spadaro *et al,* 2022). OTML is continuing studies to further improve mine waste management through to and beyond mine closure to minimise long-term environmental impacts from the project.

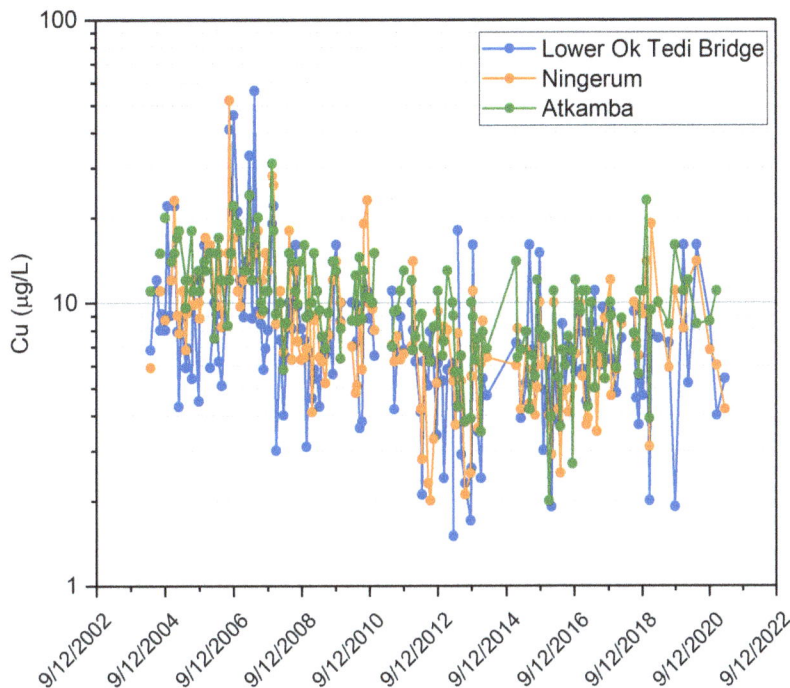

FIG 2 – Dissolved copper concentrations in the Ok Tedi between 2004 and 2021. Sampling sites include Lower Ok Tedi Bridge just downstream from the mine, Ningerum approximately halfway downstream between the mine and Bige, and Atkamba just downstream from Bige.

ACKNOWLEDGEMENTS

OTML is gratefully acknowledged for allowing the publication of this material and for their support of the studies described.

REFERENCES

Rathi, B, Schumann, R, Stewart, W, Kepe, E and Pile, J, 2022. Model-based forecasting of limestone dosing for better management of tailings NAPP values at Ok Tedi mine, in *Proceedings of the 12th International Conference on Acid Rock Drainage*, pp 767–774 (The University of Queensland: Brisbane).

Spadaro, D A, King, J, Angel, B, McKnight, K S, Adams, M S, Binet, M T, Gadd, J B, Hickey, C W and Stauber, J L, 2022. Probabilistic risk assessment of mine-derived copper in the Ok Tedi/Fly River, Papua New Guinea, *Sci Total Environ*, 810:151219.

Moving beyond Net Present Value project evaluation to achieve Environmental, Social and Governance goals for mine operations and closure

H Simpson[1] and C Gimber[2]

1. Consulting Director, ERM, Brisbane Qld 4000. Email: howard.simpson@erm.com
2. Partner, ERM, Brisbane Qld 4000. Email: chris.gimber@erm.com

ABSTRACT

The financial evaluation of mining projects drives a whole range of decision-making regarding project viability, project design, operating models and closure philosophy. Net Present Value is a financial assessment approach that is pervasive across the industry and has been widely criticised for driving short-term thinking and leading to poor long-term outcomes and future unfunded closure liabilities. But are there alternative models and what are the impediments to their implementation?

This paper aims to outline current and past mine closure practices in the mining industry and questions these outcomes given the changing shareholder investment requirement for alignment with sustainable Environmental, Social and Governance (ESG) goals. Examples are presented of closure programs planned and executed in the later stage of the life-of-mine and explores the benefits that could have been realised by integrating these into concurrent production operations. Examples of simultaneous mine development and closure execution and their benefits are presented, and their relevance to corporate investor branding.

This paper also discusses the concept of investor project ownership through to sustainable closure and outlines alternative approaches to responsible, sustainable mine closure. We explore the current mining industry divestment practices to remove closure liability from the balance sheet. These underlying financial practices and Net Present Value approach to Mine Closure and alternative approaches are discussed.

Above and beyond NPV – determining the life-of-mine from an economic perspective

C Tapia[1,2] and B Hay[3]

1. Adjunct Associate Professor, School of Minerals and Energy Resources Engineering – UNSW, Sydney NSW 2032. Email: carlos.tapia@unsw.edu.au
2. GEOVIA Senior Industry Process Consultant, Dassault Systemes, Brisbane Qld 4000. Email: carlos.tapia@3ds.com
3. GEOVIA VP Sales and Services, Dassault Systemes, Vancouver, BC V6E 3X1. Email: bradley.hay@3ds.com

INTRODUCTION

There is a close relationship between the mining industry, economic growth, and technological and social development. Mining activities benefit all economic agents, but only during a timespan determined by available reserves, primarily driven by internal factors such as production rate, technology and costs, and external factors such as prices determined by the market and taxes and royalties controlled by governments (Gordon and Tilton, 2008; Henckens, 2021; Tapia Cortez *et al*, 2018).

The intense use of Mineral Commodities (MC) needed to maintain the current lifestyle has placed ore depletion as one of the main threats to sustainable economic, social and technological development. However, ore depletion does not only arise due to the finite nature of MC, price fluctuations or technological challenges. It is also triggered by the 'economic limits of extraction', determined by the optimisation of financial factors and technical restrictions within the regulatory framework (Henckens, 2021).

Most reporting standards – if not all – define ore reserves as *'the economically mineable part of a Measured and/or Indicated Mineral Resources'* determined by the technical and economic viability of extracting, processing and commercialising MC having the NPV as the fundamental decision – making tool (JORC, 2012; Krzemień *et al*, 2016). However, is the NVP an 'economically' basis index able to capture the complexity of economic phenomena? No, it is not, and it is crucial to define what is economic and financial. Although related, they are different disciplines. Economics studies the behaviour of the different market agents, including human behaviour and policies and explains the factors involved in the scarcity (or surplus) of goods and services. Finance focuses on financial systems and managing funds considering the time, cash available, and risk. It involves banks, loans, investments, and savings and can be considered a small subset of economics (Simon, 1959; Tapia Cortez *et al*, 2018).

Given the economic nature of ore Reserves, endorsed by governments (Figure 1), NPV is an incomplete tool to determine ore Reserves as it limits the problem to simply maximising cash flows instead of optimising the economic benefits for all economic agents. In addition, using the rigid scheme of controlling taxation – unable to adapt to new market or economic conditions and systematically increasing – within the NPV framework only increases the sub-optimisation of ore reserves, mainly due to the systematic taxation rise. Short-term benefits of higher taxes and royalties are unlikely to offset the impacts of reducing exploration and project development in the long-term, which not only boosts ore depletion concerns but also reduces competitiveness, threatening economic and social welfare (Fisher, Cootner and Baily, 1972; Gordon, Bertram and Graedel, 2007; Lilford, 2017; Mardones, Silva and Martínez, 1985; Radetzki, 2009; Rodrik, 1982; Tan, 1987; World Bank, 2000).

Holistic methods and collaborative tools, able to dynamically simulate and assess how changes in the technical settings, regulatory framework (taxes and royalties), and financial factors affect the availability of ore reserves and the impact for all economic agents, should be incorporated into the evaluation process to fill the gap.

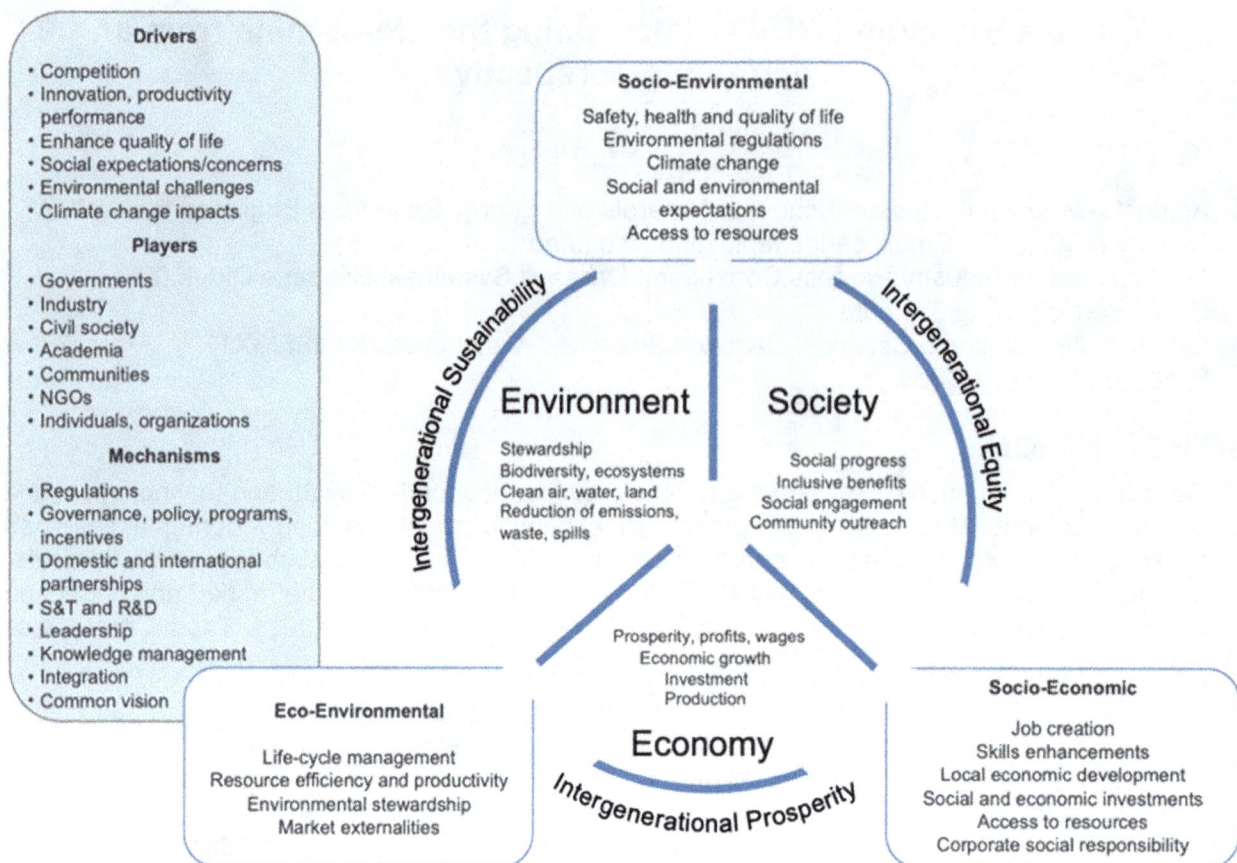

FIG 1 – Elements of a Responsible and Sustainable Approach (Government of Canada, 2013).

GOVERNMENTS, TAXES, ROYALTIES AND REGULATIONS

Most regulations aim to ensure the safe, healthy and least environmentally damaging criteria to exploit mines. However, in practice, the main target is to establish mechanisms of how governments or states will receive the financial benefits from mining activities while trying to maintain an attractive environment for investment. In contrast, governments often require sinking funds with a significant upfront deposit or regular payments to ensure enough funding for mining closure and environmental reparations once mining has ceased (Lilford, 2017). It is hard to perceive these strategies as good instruments for stimulating investment and long-run operations. In addition, mineral taxation alternatives are commonly rigid. Royalties are typically unit-, value – and profit-based (Castillo and Hancock, 2022); however, they leave no option for companies to navigate through different economic scenarios as governments have the market power to dictate the instrument for the industry unilaterally.

For investors, prices should be high enough to cover mandatory payments and operational and capital costs to maintain a sustainable industry. Thus, taxes and costs affect the industry's profitability, determine ore reserves, the feasibility of new projects and, ultimately, government income from taxes. For example, the cut-off grade is calculated pre-tax basis, while profit is calculated post-tax. Suppose taxation (royalties) is levied against the income. In that case, it must be incorporated into the cut-off grade, making it higher and forcing the sterilisation of ore reserves (Lilford, 2017).

Due to their divergent goals, companies and governments have different preferences. For example, while governments prioritise payment stability by receiving constant tax revenue (unit/value), companies may prefer profit-based taxation to avoid misrepresentations to make marginal decisions. Thus, while governments like the certainty of receiving taxation payments, companies would not be too concerned if they could mitigate the payment of taxes, royalties or both. Hence, the optimum solution lies between the two (Castillo and Hancock, 2022; Lilford, 2017). However, for that, economic factors must be taken into account and assessed with a long-term view, including inflation targets and trends, internal demand, employment rate and gross capital formation, to name a few.

THE ECONOMIC SETTING TO DEFINE ORE RESERVES

Natural resources subjects involve complex dynamic systems. Determining the technical-economic feasibility and ore reserves availability requires a deep knowledge of the technical and economic domains to understand and develop intricated engineering, financial and economic models (Castillo and Hancock, 2022; Tapia, Coulton and Saydam, 2020). However, it is not always the case.

From the technical side, engineering studies shallowly address price forecasting, labour market and salaries, which are more a collection of consensus information than an evaluation of the economic implications of projects. Sensitivity analyses commonly addressed changes in ore reserves isolated and linearly by measuring the impact on the NPV within a range of pre-established fluctuations. However, it does not link the taxation framework and other economic variables to evaluate the actual impact on the LoM and the long-term effects of ore sterilisation that, based on financial outcomes only, will likely remain economically unviable in our lifespan.

From the policymakers' side, the mining taxing scheme is commonly rigid due to its inability to choose or modify it while production and other environmental/social milestones are accomplished. In addition, taxation does not always recognise the technical nor the market differences between MC, which is essential to maintain competitive markets, promote investments and address environmental and social concerns. Increasing taxes may reduce exploration and development activities in the longer term, and the near-term gains through higher taxes and royalties are unlikely to offset longer-term welfare issues. Unfortunately, high price periods have encouraged the emergence of economically nationalist governments that have increased taxes, royalties, and nationalised assets or strategic MC (Radetzki, 2009; Slade, 2015; World Bank, 2000).

The combination of lack of economic representation and government taxation instability affects the valuation of mining projects with direct implications for both companies and local economies (Lilford, 2017):

- Reducing the LoM, employment rate and shortening the monetary capacity.

- Pressing governments to take responsibility for the early retrenched employees.

- Accelerating mining closure cost for operators.

- Early cease of taxes and royalties incomes for governments.

- Negative impact on communities and on the equity market affecting the ability to raise capital for further development.

The lack of real representativeness limits the NPV financial exercise to a pre-defined area of possible static scenarios, unable to explain the economic nature of ore reserves established by reporting standards nor their effects over time. Before establishing the taxation level, policymakers must assess whether obtaining immediate large returns or long-term employment and lasting internal demand is more critical and determine the impact of increasing taxes and the likelihood of mining companies discontinuing operations.

Figure 2 depicts how the regulations and LoM strategy affect ore reserves and financial and economic outcomes. Using gold assets as study cases shows that maintaining the ore reserves baseline reduces the financial performance, yet at a similar or slightly lower scale than reducing the life of the mine. It means sterilising ore reserves, cutting jobs and taxing income do not provide significant financial differences. It demonstrates that by working together, government and companies can find the best strategy to ensure financial performance while avoiding anticipated ore depletion is essential to ensure a sustainable level of ore reserves, avoid scarcity threats and benefit all agents.

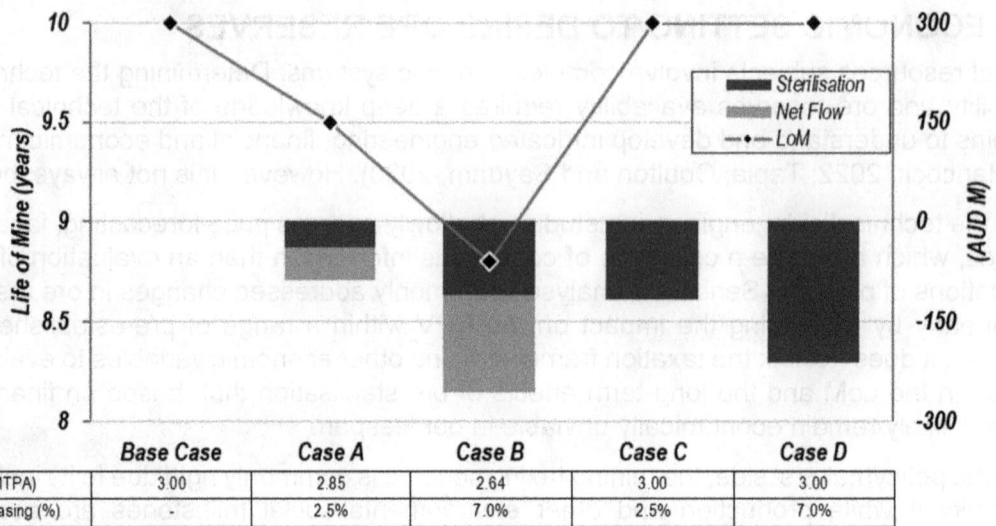

	Base Case	Case A	Case B	Case C	Case D
Prod. Rate (MTPA)	3.00	2.85	2.64	3.00	3.00
Royalty Increasing (%)	-	2.5%	7.0%	2.5%	7.0%

FIG 2 – Effects on Ore Reserves Comparing by financial flows (Adapted from Lilford, 2017).

HOLISTIC METHODS AND COLLABORATIVE TOOLS

To be sustainable, capital-intense enterprises like mining and urban planning require active interaction between social enterprise and government to face the challenge of balancing (and making co-exist) long-term strategic planning with short-term demands, but bureaucratic, organisational, planning, communication and investment attractiveness deficiencies commonly surround the process. Another obstacle to overcome is the distance between the top-down and community, generating scepticism of politics and policies and, in some cases, rejection of projects. Urban planning has overcome those limitations by replacing the short-term perspective with a long-term approach. New smart cities are built incorporating society in the decision-making process to ensure that city strategies match citizens' goals. It is obtained by providing collaborative digital platforms that allow efficient communication between society-government-enterprises that helps policymakers to plan and assess their decisions in a more informed manner. This new socio-economic model has demonstrated that society and enterprises can be part of the governance. Mining can find a reference in urban planning and replicate its practices (Angelidou, 2017; Correia, Marques and Teixeira, 2022; d'Alena, Beolchi and Paolazzi, 2018).

New technologies, such as digital twin, simulation and online enterprise communication platforms, are helping cities to solve technical, environmental, infrastructure, logistics, waste management, safety and many other problems that are also part of mining. These technologies provide the adaptivity and fast reaction needed to deal with and simulate the diversity and uncertainty features of contemporary cities and mining assets (Szpilko, 2020). An example of collaboration can be depicted in Japan, where a collaborative digital platform has been deployed and used to optimise mobility and energy consumption in a constantly growing city, where scalability was also part of the solution (Bloomberg, 2021).

Simulation tools mimic physical phenomena with high accuracy and can evaluate thousands of feasible scenarios embedded in complex linear or dynamic models, increasing the chance of finding the optimum combination that matches long-term strategy and short-term demands. In the case of mining, simulating multiple scenarios to determine the global optimum is not a trivial task as it requires the interaction of many departments and the integration of multiple data sources that make Ore Reserves optimisation a lengthy and time-consuming task so that only a few scenarios can be effectively evaluated. With this limitation, it is uncertain if the best option has been chosen and if the optimum amount of Ore Reserves will be extracted over time. Indeed, most of those few scenarios cannot simultaneously capture and effectively assess market changes like prices, interest rates, operational and investment settings or the effects and benefits of a more dynamic taxation framework that allows enterprises and government to find the more beneficial long-term strategy maximising short-term welfare for all economic agents. Figure 3 depicts the power and effectiveness of a comprehensive simulation process to determine Ore Reserves, where driving variables can be simultaneously changed to simulate even uncertain scenarios such as commodity prices, cost, and even taxation and royalties.

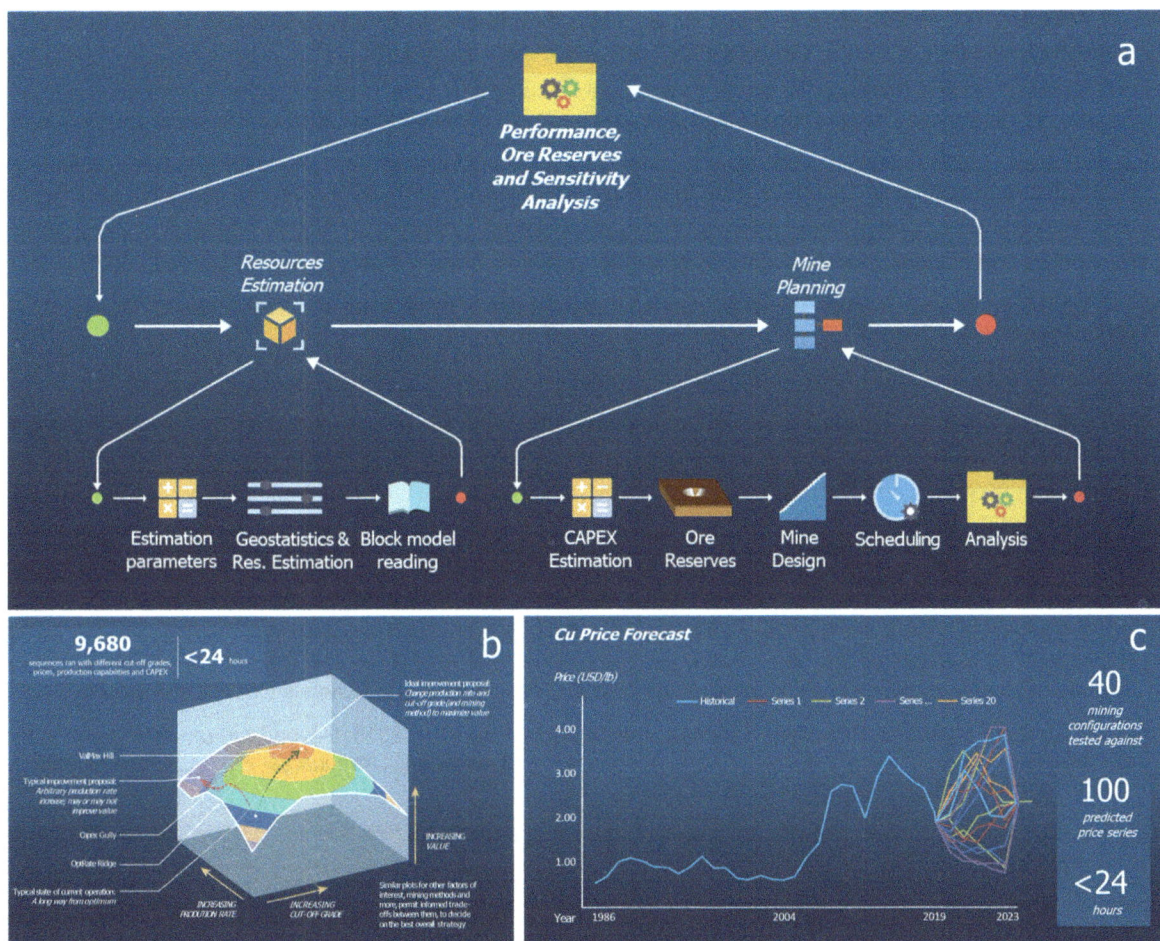

FIG 3 – Holistic Simulation to Determine Ore Reserves. Adapted from Dassault Systèmes (2022) and Hall (2014). **(a) Simulation – Process Composer:** Iteration loop running processes, automation, Design Of Experiment (DOE), optimisation and integrating all surrogate models required to evaluate each scenario (dynamic or linear models). The process can be defined as a deterministic exercise running a set of stochastic scenarios changing variables values, linking their relationship or making intermediate calculations to obtain – for example – Gaussian distributions (probabilistic models). **(b) Hill of Value:** Holistic visualisation of the current state and 15 000 scenarios comprehensively simulated to assess the potential value that can be gained due to changes in driven variables. **(c) Re-assessment of Ore Reserves level and Risks:** Incorporating economic uncertainties of commodity price by evaluating 40 mining configurations over 100 commodity price forecasting models.

In summary, holistic simulations allow the evaluation of virtually all technically, politically and economically feasible scenarios to determine Ore Reserves. Incorporating these outcomes and analysis into collaborative platforms may promote closer interaction between all economic agents as, based on tangible deterministic outcomes of almost all feasible options, they can discuss in a more informed manner how their short-term welfare support and give continuity to reach the long-term benefits established at a strategic planning level and; therefore, contribute to alleviating ore depletion concerns to maintain a sustainable economic, social and mining industry development.

REFERENCES

Angelidou, M, 2017. Smart city planning and development shortcomings, *TeMA-Journal of Land Use, Mobility and Environment*, 10(1):77–94.

Bloomberg, 2021. Dassault Systèmes and NTT Communications Announce Alliance for Sustainable Smart Cities in Japan. Available from: <https://www.bloomberg.com/press-releases/2021-12-16/dassault-syst-mes-and-ntt-communications-announce-alliance-for-sustainable-smart-cities-in-japan>

Castillo, E and Hancock, K J, 2022. Multiple streams framework and mineral royalties: The 2005 mining tax reform in Chile, *Resources Policy*, 77. Available from: <https://doi.org/10.1016/j.resourpol.2022.102722>

Correia, D, Marques, J L and Teixeira, L, 2022. City@ Path: A collaborative smart city planning and assessment tool, *International Journal of Transport Development and Integration*, 6(1):66–80.

d'Alena, M, Beolchi, S and Paolazzi, S, 2018. Civic Imagination Office as a platform to design a collaborative city, *ServDes2018: Service Design Proof of Concept, Proceedings of the ServDes 2018 Conference,* 150:645–648.

Dassault Systèmes, 2022. *Illuminate your Mine's Long-Term Value.* Available from: <https://www.3ds.com/industries/infrastructure-energy-materials/make-mine-planning-decisions-confidently-even-in-times-of-market-uncertainty>

Fisher, F M, Cootner, P H and Baily, M N, 1972. An Econometric Model of the World Copper Industry, *The Bell Journal of Economics and Management Science*, 3(2):568–609. Available from: <https://doi.org/10.2307/3003038>

Gordon, R B, Bertram, M and Graedel, T E, 2007. On the sustainability of metal supplies: A response to Tilton and Lagos, *Resources Policy*, 32(1):24–28.

Gordon, R L and Tilton, J E, 2008. Mineral economics: Overview of a discipline, *Resources Policy*, 33(1):4–11.

Government of Canada, 2013. Mining Sector Performance Report, 1998–2012, Energy and Mines Ministers' Conference August 2013.

Hall, B, 2014. *Cut-Off Grades and Optimising the Strategic Mine Plan* (The Australasian Institute of Mining and Metallurgy: Melbourne), pp 110–112.

Henckens, T, 2021. Scarce mineral resources: Extraction, consumption and limits of sustainability, *Resources, Conservation and Recycling*, 169. Available from: <https://doi.org/10.1016/j.resconrec.2021.105511>

JORC, 2012. Australasian Code for Reporting of Exploration Results, Mineral Resources and Ore Reserves (The JORC Code) [online]. Available from: <http://www.jorc.org> (The Joint Ore Reserves Committee of The Australasian Institute of Mining and Metallurgy, Australian Institute of Geoscientists and Minerals Council of Australia).

Krzemień, A, Riesgo Fernández, P, Suárez Sánchez, A and Diego Álvarez, I, 2016. Beyond the pan-European standard for reporting of exploration results, mineral resources and reserves, *Resources Policy*, 49. Available from: <https://doi.org/10.1016/j.resourpol.2016.04.008>

Lilford, E V, 2017. Quantitative impacts of royalties on mineral projects, *Resources Policy*, 53. Available from: <https://doi.org/10.1016/j.resourpol.2017.08.002>

Mardones, J, Silva, E and Martínez, C, 1985. The copper and aluminum industries, *Resources Policy*, 11(1):3–16. Available from: <https://doi.org/10.1016/0301-4207(85)90015-7>

Radetzki, M, 2009. Seven thousand years in the service of humanity—the history of copper, the red metal, *Resources Policy*, 34(4):176–184.

Rodrik, D, 1982. Managing resource dependency: The United States and Japan in the markets for copper, iron ore and bauxite, *World Development*, 10(7):541–560. Available from: <http://www.sciencedirect.com/science/article/B6VC6-45CWVFT-77/2/33ab7f808bede29867b77b98be67a378

Simon, H A, 1959. Theories of Decision-Making in Economics and Behavioral Science, *American Economic Review*, 49(3):253–283. Available from: <https://doi.org/10.1257/aer.99.1.i>

Slade, M E, 2015. The rise and fall of an industry: Entry in U.S. copper mining, 1835–1986, *Resource and Energy Economics*, 42:141–169. Available from: <https://doi.org/10.1016/j.reseneeco.2015.08.001>

Szpilko, D, 2020. Foresight as a tool for the planning and implementation of visions for smart city development, *Energies*, 13(7):1782.

Tan, C S, 1987. *An econometric analysis of the world copper market* (SCP 20; World Bank Staff Commodity Paper, Available from: <http://documents.worldbank.org/curated/en/1987/10/1555519/econometric-analysis-world-copper-market>

Tapia, C, Coulton, J and Saydam, S, 2020. Using entropy to assess dynamic behaviour of long-term copper price, *Resources Policy*, 66. Available from: <https://doi.org/10.1016/j.resourpol.2020.101597>

Tapia Cortez, C, Saydam, S, Coulton, J and Sammut, C, 2018. Alternative techniques for forecasting mineral commodity prices, *International Journal of Mining Science and Technology*, 28(2):309–322. Available from: <https://doi.org/10.1016/j.ijmst.2017.09.001>

World Bank, 2000. *Global Commodity Markets: A Comprehensive Review and Price Forecasting.* Available from: <http://www-wds.worldbank.org/servlet/WDSContentServer/IW3P/IB/2000/08/14/000094946_00080105305346/Rendered/PDF/multi_page.pdf>

Incorporating the true costs and opportunities of rehabilitation and closure in project selection, design and approvals

C Tomlin[1] and C Gimber[2]

1. Principal Consultant and Team Lead, ERM, Brisbane Qld 4000.
   Email: charissa.tomlin@erm.com
2. Partner, ERM, Brisbane Qld 4000. Email: chris.gimber@erm.com

INTRODUCTION

Historically, key closure risks have been studied during operations and then further assessed as cessation of mining approaches. Only in the past couple of decades or so has progressive rehabilitation and closure become an integral part of regulatory frameworks, sustainable mine planning and more recently, new project and development planning.

It is now broadly accepted that the best closure outcomes occur when rehabilitation and closure are properly considered upfront, during the early stages of a mine development project and well before mining occurs. Increasingly, regulatory processes are requiring more diligence on early closure planning, often as part of the environmental impact assessment (EIA) and approvals process, rather than later in dedicated closure study phases. There is an opportunity to use the information generated during this phase to make better decisions on project selection, design, mining methodology and all other facets of the mining operation.

BARRIERS TO UPTAKE

Incorporation of rehabilitation and closure risks and criteria in early project concept, feasibility and planning phases can improve decision-making to reduce whole-of-life costs and maximise Environmental, Social and Governance (ESG) benefits. Whilst some progress has been made in this area, it is yet to be fully embedded into the project selection and option selection phase. Additionally, preliminary costings associated with rehabilitation and closure are often largely underestimated. There are a number of factors leading to this:

- Understanding – The level of information available on a project increases throughout its development. Often the true risks related to closure are difficult to quantify or may not even have emerged (eg community sentiment) at the early stages of a project.

- Process – most organisations have well developed processes for making decisions and selecting projects (or project components) to be carried forward into development. However, these processes often do not give sufficient weight to the closure-related aspects of a project. Closure is often given a cursory glance, rather than a detailed review that truly affects project selection.

- Optimism – before mining occurs the project team envisages how the operation will develop and how the project will unfold, and often will incorporate best practice features with good intention, such as encapsulation of Potentially Acid Forming (PAF) materials, for example. Assumptions are also made that everything will run according to plan – waste will be properly characterised; sent to the correct location; paddock dumped and traffic compacted; and a suitable cover will be available and constructed. There are many aspects that can go wrong, but before the project starts there is a tendency to assume everything will run along the ideal path.

- Bias – at the project selection phase many of the stakeholders are motivated, either consciously or subconsciously, to get the project off the ground. Focus is usually on developing a financially viable project often with short-term objectives in mind and looking to reduce the payback period of the capital expenditure. This can come at the expense of long-term, whole of life cost analysis.

- Time frame – closure issues manifest themselves over the life of the operation. They often will not appear until later in the mine life cycle, or even after mining has ceased. For many developments, the problems are unlikely to be managed by the people making the early decisions around project selection and design, and this can lead to less responsibility and ownership from the project team.

EMBEDDING CLOSURE IN PROJECT SELECTION

The best way to overcome these barriers is to be conscious of them at the project selection phase. The biases, time frame, knowledge gaps and certainties will always be present, but with the correct processes in place, it is possible to properly embed closure into the decision-making process. This can be achieved by:

- Learning from the past – not repeating mistakes, improving practices, and sharing successes and failures (between sites and industry).

- Assembling the right teams – having truly multidisciplinary teams involved in decision-making. This should include representation by technical specialists in social and community, closure design and execution, environmental management, a range of engineering disciplines, site-based teams (where applicable), schedulers and cost control.

- Having the right processes – making sure that development teams follow a consistent process. The selection process should not bias any particular outcome.

- Costing – considering costs over the full life cycle of an operation.

One of the main tools used in option assessments is Multi-Criteria Analyses (MCAs). MCAs often touch on high-level closure considerations, but option and design development is rarely based on future, detailed progressive rehabilitation and closure requirements.

During the assessment or MCA process, project teams should be asking questions like:

- Has the project location considered proximity to sensitive receptors (environmental, social and cultural heritage) and not only construction and operational impacts, but closure and post-closure impacts and management (ie potential nuisance dust, visual amenity).

- Will any of the options result in differentiating rehabilitation outcomes (ie slope angle or length that may influence the ability of vegetation to establish on embankments)?

- Will a liner or additional seepage control measures be required?

- Will the Post-Mining Land Use (PMLU) be achieved by the final landform of each option considered and is the closure vision proposed in line with the existing site-wide closure objectives and vision?

These are just a few of the potential rehabilitation and closure considerations that should be thought about when planning new developments and selecting options. Furthermore, upfront development of detailed rehabilitation prescriptions and closure completion criteria can also reduce uncertainty with Regulator's acceptance of project approvals and minimise delays to final relinquishment.

CONCLUSION

Implementing meaningful rehabilitation and closure considerations in the options selection process beyond Net Present Value (NPV) can lead to many benefits that may result in a change to the project design (including location of infrastructure and activities), construction methodology, complexity of approvals pathway / processes and operational procedures. Additionally, determining detailed costs of the identified rehabilitation and closure activities and building these costs into capex and opex allocations at an early stage is essential. It is fundamental that these approaches are developed to ensure closure-related risks and opportunities are captured in strategic mine planning.

AUTHOR INDEX

www.ingramcontent.com/pod-product-compliance
Lightning Source LLC
Chambersburg PA
CBHW061104210326
41597CB00021B/3977